T0284764

A Tale of Two Cranes

Lessons Learned from 50 Years of the Endangered Species Act

Nathanial Gronewold

PB Prometheus Books

An imprint of Globe Pequot, the trade division of The Rowman & Littlefield Publishing Group, Inc.
4501 Forbes Blvd., Ste. 200
Lanham, MD 20706
www.rowman.com

Distributed by NATIONAL BOOK NETWORK

British Library Cataloguing in Publication Information Available

Library of Congress Cataloging-in-Publication Data

Names: Gronewold, Nathanial, 1977– author.
Title: A tale of two cranes : lessons learned from 50 years of the Endangered Species Act / Nathanial Gronewold.
Other titles: Lessons learned from 50 years of the Endangered Species Act
Description: Lanham, MD : Prometheus Books, [2023] | Includes bibliographical references and index. | Summary: "A Tale of Two Cranes will serve as a launching pad for better understanding the progress and pitfalls inherent in endangered species management, through 50 years of lessons learned since the landmark Endangered Species Act was enacted by the United States Congress in December 1973"— Provided by publisher.
Identifiers: LCCN 2022052417 (print) | LCCN 2022052418 (ebook) | ISBN 9781633887626 (cloth) | ISBN 9781633887633 (epub)
Subjects: LCSH: United States. Endangered Species Act of 1973—History. | Biodiversity conservation—United States. | Biodiversity conservation—Government policy—United States. | Cranes (Birds)—United States—Conservation. | Gronewold, Nathanial, 1977– .
Classification: LCC QH76 .G746 2023 (print) | LCC QH76 (ebook) | DDC 333.95/220973—dc23/eng/20230113
LC record available at https://lccn.loc.gov/2022052417
LC ebook record available at https://lccn.loc.gov/2022052418

Printed in India

For Atsuko

~

Contents

Introduction

Many of you will find this to be a rather unusual book to hold. For starters, this is a book about the U.S. Endangered Species Act, but the cover features a species that's not actively protected by the Endangered Species Act, at least not directly. That's on purpose, actually. Some of you will also find it rather peculiar to read, as well—many of the chapters run off topic and in strange tangents at times. That's intentional, too.

The main theme and topics of this book may seem somewhat esoteric at first glance, but in fact this book is intended for a wide audience, crane fans or not, despite my dabbling in specialized language here and there. This book is narrative in many parts, analytical in others, and oddly autobiographical throughout, but it's all very much intended. I could have written a drab technical treatise for college professors and wildlife management professionals, but that's no fun, and that's not what I've done here. There's still plenty of forage for those professionals to graze on in these pages, though. I promise. But this book isn't only for endangered species protection enthusiasts or conservation professionals—it's for everyone, regardless of professional background or level of educational attainment.

But I'll admit straightaway that this is an odd book considering the topic it tackles and the approach I took in analyzing it. Oftentimes while reading you may wonder, *what is this guy going on about*? But please soldier through these sections as best you can, because they do have a point. There's ultimately a method to the madness, I promise, though it may take time to discern it.

And to the conservation professionals and journalists who may feel offended by much of what's written in the following pages, I ask that you all soldier through those sections as well, bravely and without regret. We may have differences of opinion, but we can certainly disagree without being disagreeable. I can read news articles or books containing content that I vehemently disagree with without feeling personally offended or attacked by any of it. In parts of this book I criticize professional practices, but I never attack anyone personally. It's a necessary balance that I take given the severity of the world's extinction crisis.

Finally, a brief word of heartfelt thanks to the many colleagues, associates, friends, acquaintances, and loved ones who helped me get this book to print.

I couldn't have written this book and the facts and figures contained within it were it not for the enormous well of support I was able to draw from. To my past and present graduate school advisers in Texas and Hokkaido, thank you for your support. Thank you to the professionals in Akan, Tsurui, Kushiro, Sapporo, Texas, Aransas, Colorado, and Washington, D.C., who were kind enough to share their time and expertise with me over the past several years in person, by phone, and via email. Most of the encounters and interviews retold in this book occurred in my beat reporting days. I gained more from these women and men than I could ever imagine, and I'm grateful that they agreed to lend their words and wisdom to my writing. Thank you all. Thank you to my colleagues at the small university in Tsukuba where I launched my academic career. Special thanks to Jake and everyone at Prometheus for waiting patiently for my first draft and for the subsequent aid they extended in seeing this book to print. Thank you to my copy editor for greatly improving this book. And a big thank-you is in order to my wife, who encouraged me to pursue my research and dreams, and to my family, who've been a solid wind at my back.

And thank you, reader, for finding this book intriguing enough to pick up and read. You're the real power and passion behind *A Tale of Two Cranes: Lessons Learned from 50 Years of the Endangered Species Act*. You're the reason why I wrote this book, my second to make it to print, and I sincerely hope you enjoy it.

CHAPTER ONE

~

No Fish, No Fowl

Between December 1965 and August 1966, a Japanese science-fiction journal called *SF Magazine* published a serialized tale by the author Mitsuse Ryu. His story was titled *10 Billion Days and 100 Billion Nights*, as translated to English, of course. The adventure Mitsuse delivered to the magazine's audience of sci-fi enthusiasts that year later inspired a manga series and eventually made its way into published novel form. I happen to have both the Japanese and English versions of this fantastic novel in my possession.

I mention all this here because Mitsuse's unique tale begins with a poem about the constant ebb and flow of subtle yet permanent changes that occur on Earth endlessly, unceasingly, for what seems to us like an eternity. The Japanese version of the poem is lovely in its own right, but it's not quite as eloquent as the English translation, in my humble opinion.

One bit, as translated by Alexander O. Smith and Elye J. Alexander, stands out for me. Allow me to share that portion of Mitsuse's translated poem with you here:

> All are absorbed and reduced into individual molecules, tiny motes that show no hint of their vast history. In the bottomless sediments, only a vague memory remains.
>
> The sea: it contains within itself the long, long story of time, a perpetual record of shapes that will never be seen again.

The second half, in particular, resonates for me: "a perpetual record of shapes that will never be seen again." Just think about those words for a second and what they're trying to convey. Let them resonate in your head for a little while.

In a nutshell, Mitsuse's poem is about how the dust of everything in existence on Earth eventually ends up at the bottom of the ocean. You and I and all of us, that's where we're headed: the sea. It's fitting, considering how science is strongly convinced that it was from the sea that all life on Earth emerged in the first place. Endlessly roiling and churning, never ceasing in its motion, gradually and relentlessly, the sea eats away at the coasts and mountains, valleys and continents, over long eons of time, swallowing the dust and ashes and molecules and fossils of countless millions of organisms large and small that once roamed this planet just as you and I do now. Only vague shadows of their former selves remain, no pictures or hard memories, and only if we're lucky. It is exceedingly rare for any living creature to ever become a fossil. That we have so many fossils of long-extinct living things for us to research, marvel at, and display to the curious public in our museums speaks to the tremendous multitude of living creatures that have ever lived and walked on our world.

For the vast majority of formerly living things, their shapes will never be seen by anyone or anything ever again. Not even their footprints, nor the shadows of their skeletons. They will remain hidden from us permanently, for eternity, fantastic creatures whose remains never chanced to fossilize in any way whatsoever. The only memory of them left, if you can call it that, resides in the molecules of water that pass through our bodies or in the atoms of the dust that cling to our nostrils. All of it eventually finds its way to the sea, "time's closest confidant" as Mitsuse so poetically puts it. He had it right: the ocean contains within its immense belly "a perpetual record of shapes that will never be seen again."

The oceans are slowly but surely eating away at most of the Hawaiian Islands, England, North America, South America, Asia, Australia, New Zealand, and pretty much any bit of geology that isn't currently being built or added to through uplift or volcanic activity or shored up (temporarily) by our clever human engineering. But that line in Mitsuse's poem isn't pointing only to geology—of the past mountains and valleys, canyons and glaciers, rivers and lakes, volcanoes and islands and more that will never be seen by us or any other living creature again, ever. His poem holds a deeper meaning, I like to think.

The Caribbean monk seal (2008). The Yangtze River dolphin (2007). Lienard's giant gecko (1996). The Japanese river otter (2012). The Carolina parakeet (2016). The Pinta giant tortoise (2015). The Floreana giant tortoise (2017). The Navassa Island rhinoceros iguana (2010). Schomburgk's deer

(2014). The Utah Lake sculpin (2011). The Christmas Island pipistrelle (2016). The laughing owl (2016). The large sloth lemur (2008). The Charco azul pupfish (2018). The Guam flying fox (2019). These are some other shapes that neither you nor I nor anyone or anything else will ever chance to see again, if we were ever lucky enough to have caught glimpses of any of them at all before.

Sure, you may chance to witness these vanished animals in photographs or perhaps even view a long-dead and stuffed specimen in a museum or university lab somewhere sometime, should you seek to do so, but this is not what I mean. These animals were once moving, breathing, living creatures with which we shared this planet. Their shapes were endlessly changing, in terms of their morphology and in the motions and sounds they made or in the tracks and footprints they left in their wake. Perhaps even in the eggshells they left behind. They were born or hatched then grew. They moved, hunted or foraged, or did both. They slept when they were exhausted. They ran from threats and kept a wary distance from curious onlookers. They cried. Then they died. And then eventually all of them died, every last living member of their species. They're now shapes you and I will never see again, invisible molecules either headed for the sea or already there.

The list of creatures I mentioned includes animals recently declared extinct by the International Union for Conservation of Nature, or IUCN, a global team of scientists and conservationists whose mission is to catalog and preserve what remains of Earth's quickly vanishing biodiversity. The year each species was last assessed and either declared or confirmed as extinct by IUCN members is included in parentheses next to the species' names. I failed to include a comprehensive list of all recently declared extinct animals and neglected to mention any recently declared extinct plants at all. The catalog of extinct plants is probably far larger. Extinct animals are more memorable, of course, because they move and leap and fly and do other fantastic things. Or at least, they used to, which is why there is so much more written about them. Though we shouldn't forget about the plants, this book is mainly about the animals, as such books usually are. Apologies in advance for this.

The Carolina parakeet once flew over much of the U.S. Southeast and parts of the Midwest. It was probably poisonous; not via its bite or scratch but to anything that might consume one. But this evolved defensive mechanism apparently didn't save it—this species hasn't been seen for decades.

The two species of giant tortoise listed earlier recently disappeared from the Galapagos Islands in Ecuador. They're gone for good, too, I'm afraid, and others may soon follow, since Galapagos tortoises generally are endemic only to certain islands in that archipelago. The Yangtze River dolphin was very recently declared extinct—try as they might, the few Chinese scientists who insist that this species is still probably around someplace haven't managed to spot evidence of even a single surviving individual anywhere. I once traveled down a lengthy stretch of the Yangtze River myself, and I don't recall seeing any animals in that waterway, let alone something the size of a dolphin. And that vanished species may soon have more company in its category: every remaining species of freshwater dolphin throughout the world is on the brink of extinction, unfortunately, according to a 2020 IUCN report.

This brings to my mind another cetacean, though not the freshwater kind.

The vaquita is a small mouse-faced species of porpoise that exists only in the northern stretches of the Sea of Cortez in western Mexico. Adult vaquitas grow to about the size of a large dog. There are fewer than a dozen of them left, if that, and there is a very high probability that this animal will soon cease to exist entirely, perhaps even by the time I finish writing this book. The vaquita is disappearing for the stupidest of reasons: Mexican fishermen casting huge gillnets wherever they please in the hopes of depleting another critically endangered species, a rare type of fish called the totoaba, all because far too many credulous consumers in East Asia have fooled themselves into believing that eating totoaba swim bladders will give them magical powers of longevity. Wayward vaquitas become entangled in these gillnets and drown. The vaquita is quite possibly the most critically endangered animal on the planet, and scientists don't know if they'll be able to save the species at all. The Mexican government has shown very little interest in rescuing and conserving this species until very recently, and past attempts by scientists to bring vaquitas into safe captivity for possible captive breeding ended in disaster.

How about some more famous examples?

The thylacine, otherwise known as the Tasmanian tiger, vanished from this planet in the early 1900s (though there have been some unconfirmed sightings, including a report given as recently as the 1980s, the species is almost certainly gone for good). Another famous namesake of that Australian island, the Tasmanian devil, may one day soon join the thylacine into oblivion, though a team of conservationists is determined not to let this happen. Tasmanian devils are being ravaged by a type of highly contagious face cancer of unknown origins (but probably somehow human caused, if we're being honest with ourselves). One strategy proposed by conservationists to save this species is to actually allow the Tasmanian devil to become extinct in the wild while surviving only in zoos; it would be a desperate gambit, but it could end the contagion.

The passenger pigeon once swarmed the skies throughout eastern North America in the millions, in flocks large enough to blot out the sun. Our forebears killed them all, every single last one of them. The dodo, an awkward-looking flightless bird once native to the remote island of Mauritius, disappeared in probably less than 90 years from the time any human first laid eyes on one. You see, dodos were easy to kill, so early explorers decided to kill all of them. Then there's the great auk, which suffered the same fate as the dodo. This beautiful flightless bird was a skilled swimmer that happily graced high ledges and bluffs of northern latitude islands. Unfortunately, I'll never get to see one except in artists' renderings or perhaps stuffed in a museum somewhere. The great auk went extinct sometime before the Crimean War, again, thanks to humans. These are all fascinating, inspiring shapes that none of us will ever chance to see again—past peoples have robbed you and me and everyone else alive today of that opportunity.

Do you like wolves? I love them, and the late American novelist Jack London apparently had a soft spot for them, as well. I once visited a wolf sanctuary in New Jersey, the Lakota Wolf Preserve near the town of Columbia and very close to the border with Pennsylvania. Sure, the Garden State seems like an odd place for a wolf sanctuary, but wolves once thrived throughout New Jersey, and I think this particular sanctuary is a must-see for anyone interested in experiencing these beautifully haunting creatures and their eerie howls up close. Not just in New Jersey—wolves were once ubiquitous throughout North America, after all, so it's only at places like this where one can get the closest sense of what past evenings in the American wilderness might have sounded like. London would've loved this wolf sanctuary, but perhaps he and I are in the minority here.

Wolves are now far too rare in the United States, found only in sanctuaries like this one or in protected zones where federal officials can barely keep the locals from shooting them all. It's much the same story elsewhere. A past Japanese government, with the help of an American consultant, of course, very deliberately and methodically orchestrated the total eradication of the Hokkaido wolf from our planet in the late 1800s. Why? To make room for cattle, naturally. There has been some talk of reintroducing rare wolf specimens from the Asian mainland back to Hokkaido, but this remains just talk; it hasn't gone much of anywhere, since modern conservationists in Japan are only too acutely aware of the trial and error—often failing fits and starts—involved in reintroducing wolves to much of the U.S. Lower 48.

It's not just Hokkaido where something special was lost. The midnight howling of wolves hasn't been heard in any other part of Japan for more than 100 years. There are still rumors and even news reports of possible wolf sightings in rural parts of Honshu, the largest of the Japanese islands, but I liken these to Sasquatch sightings in the Pacific Northwest. Japan today is home to bears, monkeys, cranes, and something that looks like a cross between a deer and a goat—a wondrous variety of amazing wildlife. The nation is considered a biodiversity hot spot by the IUCN and rightly so—a host of oddities can be found on those islands, including one of the world's largest salamander species (truly massive) and one of the world's largest species of deer, the Ezo sika deer of Hokkaido. But not wolves. Not anymore.

In fact, today wolves are rarely encountered anywhere in the world where large human population centers are found. That's because people living 100 to 200 years ago considered them to be useless pests, like rats or termites, only good when dead. Many still hold this attitude—if America's cattle ranchers and their allies in Congress had their way, you and I would probably be unable to encounter wolves anywhere outside Alaska or perhaps Yellowstone National Park (if even in these places) or private sanctuaries like the one in New Jersey. Most of us have seen wolves only on television, and there are many powerful individuals in and out of the U.S. government who would like to keep it that way. Luckily, wolves likely won't go completely extinct anytime soon, but they came pretty close.

Let's go even further back in time.

Giant sloths. Saber-toothed tigers. Woolly mammoths. Mammoths of all sorts, in fact, both woolly and decidedly less so. Cave bears. The long-horned giant bison. The moa of New Zealand, a massive flightless bird that could kill you easily with one kick. Haast's eagle, the largest species of eagle to ever live, once hunted the massive moa and would've had little trouble lifting a human adolescent off the ground. Argentina was once home to a kind of giant armadillo that grew to the size of a Volkswagen Beetle or even larger. Chinese scientists very recently unearthed evidence of a species of hornless rhinoceros that grew to a size greater than that of elephants. They once roamed vast swaths of Asia before disappearing for good—the elephant-sized rhinos, that is, not the Chinese scientists, who are thankfully still with us. A long, long time ago, Australia was home to herds of massive snub-nosed kangaroos standing as tall as NBA basketball players. Shaquille O'Neal would feel intimidated if he ever met one, but he doesn't have to worry, because this giant kangaroo species isn't around anymore, anywhere.

Most of the animals mentioned in the prior paragraph were lost to something paleontologists call the Quaternary extinction event, a slow yet brutal cataclysm that occurred roughly between 130,000 and 8,000 years ago, coincidentally the same time period during which humans were busy spreading themselves far beyond our birthplace in Africa and colonizing the globe, robbing the world of a rich variety of biodiversity and megafauna in the process. It took guns to rid the world of passenger pigeons and thylacines. Our prehistoric ancestors exterminated some of the largest species of mammals to have ever walked Earth, using only clubs, sticks, and sharpened stones. Impressive? I suppose one could say that. But it's also staggeringly depressing to think about, in my opinion.

There's a common denominator to extinctions that occurred in the Quaternary and in our more modern times: humans.

That's right: us. Do you want to know why we can see glimpses of these fantastic creatures only in books or museums? Go look in the mirror, and you'll have your answer. All the animals mentioned here are but a small sampling of a multitude of different kinds of animals and plants that are now gone for good, a perpetual record of shapes that will never be seen again, and we're mostly to blame.

Yes, extinction is natural, indeed, inevitable. Millions of past species have been driven to extinction without our input whatsoever, and this process will continue long after we are gone. That's right—humans will become extinct one day, too. It's inevitable, I believe. But the rate of extinctions occurring on Earth since humans entered the picture has become rapid, expansive, and unprecedented.

Today in our modern times, that rate of extinctions is now accelerating, running at least 1,000 times faster than nature would have it, according to IUCN. This means that there are scores of creatures alive today that will cease to exist entirely within our lifetimes. The vaquita is but one probable example. I could write a separate book describing all the other species likely to vanish within a generation or two.

There already have been at least five past mass-extinction events recorded in the fossil record. We now appear to be amid a sixth such cataclysm. And though there is some debate about what led to the prior five mass-extinction events, it is all too obvious what is to blame for the sixth.

Us.

In Colorado, the Denver Museum of Natural History was a favorite school field trip destination of mine and my classmates back when I was a kid. I probably visited that museum on several school field trips. The museum later changed its name to the Denver Museum of Nature and Science, a new name that I don't particularly care for, but then again, I don't get to make these decisions. I hope not too much else has changed there. I haven't been to this museum in something like forever, but I still remember it well from past visits during my childhood. It of course housed the obligatory dinosaur bones and fossils, but I also remember displays of more normal-looking mammals alongside representations of pre-Neolithic hunters stalking them. I didn't recognize any of the animals on display because they no longer exist, but these renderings were instantly recognizable to me as animals of Earth and not of life on some fictional alien world. They were quite obviously real animals that once roamed the plains of eastern Colorado a long, long time ago. I remember thinking how wonderful it would have been to see these creatures alive and in person, instead of immobile as stuffed dolls. I'll never have a chance because of those pre-Neolithic hunters represented alongside them.

When it comes to our relations with our cousins in the animal kingdom, we modern humans are different from our hunter-gatherer ancestors in one very important way.

Prehistoric hunter-gatherers never once concerned themselves with the continued existence of their prey nor of the possibility that future generations may wish to experience and enjoy these animals as they did. The very concept of sustainability was alien to those people. They hunted with abandon, and when they couldn't find their favorite prey one day, they simply moved on to some other hapless species. Quite frankly, the earliest humans didn't care a whit about the continued existence of the other members of the animal kingdom with whom they shared the planet. This was true anywhere in the world. Wherever you are now reading this, know that your part of the world once crawled with a wide variety of amazing animals, often massive, many of which were eventually hunted to utter and complete nonexistence. Those ancient hunters spent no time whatsoever concerning themselves with what you and I might one day think about all this or how we might view their past actions in our present.

But something has changed only very recently, at least in terms of the span of human history.

For sure, the human ancestors who sent the mastodons and saber-toothed tigers and other fantastic beasts into oblivion didn't care at all about what they were doing. None of them spent any time debating how to sustainably harvest New Zealand's moa to ensure future generations had access to these giant flightless birds or whether the rate of human culling of giant sloths was a tad excessive and needed to be dialed back a bit. After the last woolly mammoth was slaughtered, do you think the hunters responsible for that milestone paused to mourn and reflect on this sad fact, lamenting *what have we done?* and deeply remorseful for completely ending the existence of another form or life on Earth? Of course not. Like other animals, our ancestors paid zero attention to the potential extinction of other organisms. They were completely unaware that their actions were causing scores of fantastic animals to vanish from the planet entirely. Things went on like this until around the early 1900s. That's when something changed in us—within our species.

That's about the time when people like you and I began to take notice of how quickly many of our neighbors in the animal kingdom seemed to be disappearing all around us and that we were to blame for these disappearances. Then they envisioned ways to legislate changes and to take other actions that might help prevent the complete human-driven annihilation of plants and animals, large and small.

That's the key difference today. We care. Humans care. When the vaquita is gone, we will mourn. Some of us working desperately to save the species may even cry.

Human-driven extinctions continue, for sure, but now at least attempts are made to stop this process. Today humans continually debate how best to protect other animals from extinction, even though these other animals undoubtedly don't care at all about what happens to us. We write books, like this one. Others read books like this and ponder these questions or comb through the academic literature on endangered species management in search of ideas or inspiration about how best to defend their own favorite plants or animals. Legislation is passed. Hell, entire international treaties are drafted, signed, and ratified by dozens of nations. And, of course, courts everywhere hear cases concerning the necessity of respecting the right of continued existence of other life-forms. When other life is made extinct—or threatened with extinction—we now care. We pause to think of ways to prevent these extinctions. That's the difference today.

A student of mine once asked me an interesting question: if a species is going extinct and we know that species isn't particularly central or critical to the health of an ecosystem, then why go to all the trouble of trying to save that species in the first place? This was his challenge to me. "Extinction is inevitable, right?" he said. Millions of species have come and gone, and the planet is still spinning. Considering this, why try to rescue a particular species from oblivion if we know that an ecosystem doesn't really need that species to continue thriving? This was essentially his question to me.

I first tried to answer him by explaining how we often don't know all the functions and benefits a particular species provides to an ecosystem, but this answer didn't satisfy him.

I then later suggested that protecting a species is often a good excuse for us to protect an ecosystem in tandem. Although perhaps some habitat could continue as it was without that species, that species certainly cannot survive without its habitat, so endangered species conservation becomes a good excuse to preserve and rehabilitate nature in general. Again, I received a look of dissatisfaction on his face.

After a night of reflection, I finally gave him a more honest answer: we seek to protect endangered species because we care about and like these animals (and plants), and we would feel guilty and sad if we failed to do so, especially with the knowledge that we are the cause of their extinction. In other words, we value biodiversity for its own sake, and we look back in horror at the mass slaughter inflicted by our ancestors and aspire to be better than them. My nieces may spend their adult lives in a world where the vaquita is but a memory. This fact bothers me greatly, so I would like someone to prevent this from happening, even though I may never see a vaquita in person myself whether they survive or not.

That's the honest answer. Why do we try to save species from oblivion? Because now, we care. We give a damn. It's a start.

Allow me a brief introduction and a disclaimer.

I'm not a professional ecologist or wildlife rehabilitation expert, though I am trained academically in environmental studies and ecosystem management.

I lecture at a university. My specialties are journalism, international affairs, and environmental studies. I explore how global society tries to manage its environment and the natural resources bestowed by it, either through the United Nations and its treaties and institutions or through unilateral vehicles like the Endangered Species Act. I've taught courses on the United Nations, global environmental management, ecosystem science fundamentals, and geography. By the time you read this, I will have obtained my PhD in environmental science with a focus on global environmental management, and my role in academia will have shifted to journalism instruction. Many moons ago, I worked as a private contractor in Tokyo handling media and communications work for the Japanese government. For most of my career, I've been a journalist with an emphasis on environmental and energy matters. I've spent more than seven years reporting from UN headquarters in New York. I used to cover the oil and gas industry from Texas. I closed my time as a full-time staff reporter covering the Asia-Pacific.

I'm still a journalist. I'm now the editor-in-chief of PublicParks.org, a new site launching shortly before or after this book makes print. PublicParks.org aims to be a center for news and conversations regarding the science, ecology, public policy, and management of the world's public lands and waters—national, state, and provincial parks, wildlife refuges, management zones, preserves, and protected public areas of all sorts. This includes, of course, the plants and animals that survive on these protected public parcels.

There's a reason why I'm sharing all of this. I'm good at gathering information, analyzing it, comparing facts and figures critically side by side, and then sharing my findings and some new insights in an informative but digestible manner to a broader audience of hopefully interested readers. But I'm not necessarily an expert on endangered species management. Many of you reading this are, so try keeping that in mind as you read further, since I'll probably make points or express opinions with which you vehemently disagree. It's important that these readers know where I'm coming from.

This book is about the Endangered Species Act: how it changed the world and, to some degree, how it hasn't. This book is analytical, critical, and academic. But you'll also find this read to be strangely autobiographical in parts.

This book is one outcome of an ongoing transition in my life. I hope you enjoy this part of the ride. Though I'm teaching at a university while researching environmental policy and diplomacy, I'm also writing books for readers like you. This is my most challenging writing project to date. I didn't set out on this task intending to write a bland, technocratic subject textbook that would make your eyes glaze over as you struggle to keep your head up. Nor does this book read like an extended version of an academic research paper appearing in a journal, though it does feature original peer-reviewed research I've undertaken and published elsewhere. In fact, my doctoral dissertation is contained in these pages, though hopefully you won't notice. However, parts of this book do get technical and a few chapters may be difficult for the average reader to slog through, but please slog through them anyway. You may even find some sections needlessly detailed and even repetitive. It's all intentional, just as I intended to illustrate the cover of a book on the Endangered Species Act with a species that's not actually protected by the act (at least not directly). There's a point to it all. As I take you through this story of two crane species, I am building up to a grand idea or proposal that could be considered controversial. Still, I've tried to make this exercise informative, educational, hopefully inspiring, accessible to a wide swath of readers, and entertaining at the same time. That's a tall order for a book of this sort. Let's hope I didn't screw it up, because the subject is of the utmost importance.

Earth's species, their potential extinctions, and how best to prevent said extinctions are topics that thousands of researchers far more experienced and knowledgeable than I have wrestled with for years, and no clear, definitive answers have yet to emerge. There are generally agreed-upon best practices, favored conservation policies, and fairly well-laid-out methods for assessing the relative health of a species and any risk of extinction it faces or lack thereof. Yet there is no clear road map for saving a species from oblivion. "Do A, B, and C, and your species will survive"—sorry, no such instructional manual exists, and I make no attempt to provide you with one in these pages. I do, however, offer up more interesting questions and valuable food for thought as we explore the U.S. Endangered Species Act (ESA) and endangered species management in general and consider the ESA's profound legacy. In the process, we look at where the field of endangered species protection stands today throughout the world.

As I write this, I'm sitting in my office at a university in Japan pondering how best to begin this exploration of the ESA and 50 years' worth of concerted and focused government attempts—through policies, pronouncements, executive orders, legal wrangling, fines, jail terms, and international treaties—to slow and hopefully halt the quickening rate of species' extinctions on Earth, possibly putting the brakes on the sixth mass extinction that is now underway, an extension of the Quaternary extinction event. *Endangered Species Management: How Are Things Going?* could easily be the title of this book. That line actually sums up these pages quite nicely, I think. Here's another alternative title—*The ESA at 50: What Do We Think, Folks?* Is all well and good on this much messier, more crowded, and far more confusing real-life version of Noah's Ark? Or could we change things up a bit, hopefully for the better?

Let's find out.

Yup, just 50 or so years of this going on in any serious, focused way—that's it.

Though our kind has been roaming this sphere for at least 300,000 years or so, killing and eating pretty much every kind of animal and edible plant possible in the interim, modern Homo sapiens have been taking concerted, well-organized, and deliberate cracks at preserving biodiversity only for the past 50 or 60 or so years of our entire long, long story. Yes, wildlife laws and treaties existed far earlier than this, but it was in 1973 when lawmakers in the United States gathered together and endeavored to craft a powerful piece of legislation that would end up changing the world, though at the time all they were aiming to do was ensure animals like the bald eagle, California condor, whooping crane, grizzly bear, Florida panther, manatee, and more didn't end up as mere fables in children's fantasy stories, like dragons or Sasquatch or Japan's wolves.

They achieved far more than that, I believe.

The U.S. Endangered Species Act was a pathbreaker. Other nations quickly followed the trail it blazed. In fact, the Endangered Species Act today serves as a template for a host of other similar conservation acts adopted by governments everywhere, as I later discuss in these pages.

This all comes too late for the passenger pigeon, Tasmanian tiger, great auk, and Hokkaido wolf, of course, but perhaps it isn't too late to save endangered cranes, beautiful birds of prey, whales, and much more; at least, that's the idea. Not all of these efforts were or are successful, mind you, and indeed some have proven to be colossal failures, but a wave of new species-centered legislation and international agreements were enacted in the wake of the 1973 Endangered Species Act, arguably marking the height of the worldwide environmental movement.

All this legislative, legal, scientific, and conservation activity represents a significant turning point in the history of the relationship between humans and the other living organisms on Earth, plants and animals.

For millennia, humans valued other animals only as sources of food and resources: hides for clothing, bones for tools and weapons, and other such uses. Animals were things to be captured or killed as quickly as possible and then exploited, of value only for what they could offer us in terms of immediate human welfare, living or—most of the time—dead. But that was then.

Today, we humans value wildlife far differently than in the past. Too often we still place ourselves at the top or center of a values hierarchy, but I think most of us today would agree that the animals and plants we share this planet with are valuable in their own right and that it's worthwhile to ensure that they continue to exist, even if that might seem a hopeless task at times. The enactment of the 1973 Endangered Species Act sort of clinches that moment in history when humanity finally accepted that various forms of wildlife hold intrinsic value beyond what we ascribe to them. It marked an important turning point in human history.

This makes for a terrible segue. I'm sorry, but as I noted earlier, I'm finding myself at my own turning point in my life at the moment.

As I write these words, it's been almost two years since I left a long career in international journalism for the greener acres of academia. Well, I haven't left the field completely, as I noted earlier. I still edit and report for nonprofits while leading PublicParks.org, for example. And to be a good educator, I try my best to keep up with the latest developments in the media industry, as depressing as those details can too often be. Prior to donning these new hats, I was a beat reporter for almost two decades. I truly enjoyed the vast majority of those years—it was an amazing time. I treasure each and every moment, never regretting the experience, and I'm genuinely grateful for the opportunities afforded to me during my reporting career and to the people who gifted me with the opportunities I've had. If any of them are reading these words now, I'd like to say to them thank you so much for all you've done for me.

My successes in the field were not destined nor guaranteed, and I had plenty of help along the way. I skipped journalism school entirely and taught myself the craft, ultimately reporting from ten countries on five continents and winning seven writing awards and honors, more than any other single employee in any news organization I've written for. They may not be major awards, true, but I'm rather proud of them still. I've enjoyed loads of memorable, exciting, educational, and sometimes hair-raising experiences to reflect on, far too many to mention all of them here, certainly. Along the way, I've met and spoken with scores of interesting and amazing individuals and took something with me from each and every one of these encounters. Indeed, those years I will treasure for the rest of my days.

And what of those final years? Well, quite frankly, I'd like to forget my final two years as a beat reporter ever happened.

Though it may seem cliché to say by the time you read this, Trump derangement syndrome is real and unfortunately infects countless journalists, newspaper, television, and online media professionals alike, women and men whom I once admired and looked up to. It was a sad thing to witness in real time. Some of you out there might know what I'm talking about, but many of you don't, I'm sure.

Long story short, the outcome of the 2016 U.S. presidential election took a serious psychological toll on most American reporters. Not all of them, of course, but certainly a majority. Though I long suspected that the celebrity real estate tycoon could pull off a November win that year and argued with plenty of my colleagues and friends regarding this possibility, almost no one else employed in the media industry in the United States ever considered this a real possibility. So, naturally, the news came as a profound shock to all of them. Which normally would be fine. I take no side in this political debate, but I can imagine how Hillary Clinton's defeat must have come as a terrible disappointment to her most fervent supporters.

But journalists are supposed to be capable of removing themselves from all that and simply serve the public as they're supposed to, by keeping the public informed of facts and events in as fair and accurate a manner as possible. The trouble is, the vast majority of news writers and reporters coped with Trump's unexpected victory in the worst possible way imaginable. They became embroiled in a mass delusion, collectively building within their minds an alternate reality that, to them, was the only one possible that could explain what had happened and why they never saw it coming.

You will recall, dear reader, that prior to all this, the voting public of the United States happily elected Barack Hussein Obama to the presidency twice, with sizable majority support from a diverse electorate both times. That same electorate later gave the job to reality TV star Trump. That's just the way it is. There is no contradiction here whatsoever. This was the same country, but for far too many American journalists, U.S. history began in November 2016, and all prior years no longer exist. The TV celeb won, and suddenly the Nazis were back and the Russians controlled the U.S. government, according to the U.S. media. Conspiracy theories thrived—they imagined secret hand signals being flashed on television game shows and interpreted other innocent gestures or turns of phrase as hiding ulterior, darker meanings. Half the country was written off by reporters as irredeemable and unworthy of any degree of empathy, while the news media threw fairness and accuracy out the window.

We were (are?) treated to a constant deluge of headlines that would embarrass the *Weekly World News* tabloid or even the *Onion*; only now they emblazon the pages of the most widely read newspapers and websites in the country. Once-venerated institutions like the *Washington Post* were transformed into partisan-politics-driven clickbait factories. The line between opinion and fact was largely erased and the industry's "fact checking" morphed into highly opinionated hectoring riddled with factual errors of its own. Top editors and publishers of the largest circulating coastal dailies embraced this madness and never once raised objections to any of it. All this over one election that didn't go quite as they'd hoped. The human mind truly is a fragile thing.

This seismic attitude shift in U.S. mass media essentially ripped the nation in half, forming two opposing factions or tribes. Attitudes hardened, and every minor dispute or difference of opinion became inflated into a life-and-death struggle for the soul of the country. Nothing could get done in Washington, at least not in a bipartisan way. Legislation stopped fixing America's most vexing problems, especially global warming but including its struggling biodiversity. I doubt much has improved by the time you read this. Hopefully, I'm proven wrong here.

For a time, I managed to keep my head down and avoided the worst of all this, but of course, that didn't last. The 2016 election transformed the American press corps into something more closely resembling the Borg, that relentless, all-consuming groupthink colony of cyborgs from the *Star Trek* science-fiction television series. On the TV series, the Borg colony demands complete conformity and is utterly inhospitable to independent thought or dissenting perspectives of any sort. In our reality—and during my final months as a beat reporter—the media insisted that virtually any and all reporting must tie directly to their ultimate nemesis and show him in the worst possible light imaginable. The world changed in November 2016, I was told, and I had to change with it—those were my precise orders. I'm not exaggerating, unfortunately.* The *Columbia Journalism Review*, for instance, couldn't help but notice all of this as it was occurring, and that outlet stands out as one of the few media institutions brave enough to highlight the press corps' scientifically proven obsession with this single individual and to point out how problematic it is. "Members of the press need to rethink their instinct to write endless Trump stories," *CJR* strongly advised in November 2019.[1] Of course, my former colleagues didn't take this advice.

At any rate, I went from honest admiration to feeling sorry for once-serious, formerly curious, previously competent media professionals as I was treated to daily lectures on the virtues of clickbait, rage-porn, and pandering to the audience, or on the inherent civil duty reporters had to fight the "other side" and not simply report the news, even if it meant lying to our readers and viewers. I'm venting here, of course, and I'm perhaps being too harsh on the news media. That industry still contains scores of competent professionals who know precisely what they're doing and how to best earn audience trust. But as I write these words, I can honestly say that the competent news reporters are unfortunately not a majority today, and they certainly weren't back when I began to make my career shift.

Here's a tip from a former award-winning industry pro to any current or aspiring journos who may be reading this now: when every story is about Trump, really none of them are. Rather, they become about you, the writer. Or, to put it more precisely, these obsessive reports become not about the subject of your internal angst and frustration but rather about your dangerous infatuation with this single individual, one whose story ultimately will register as just another blip in American history, a future Warren Harding.

There are other even bigger stories to cover and care about, stories with far greater and more significant lasting consequences. Climate change,

---

* By some estimates, Trump received triple the news coverage of any sitting U.S. president in history during his time in office; not 30 percent more, but three times more stories written about him compared to Obama, Bush, and the others. This was the picture when I was told "we need more stories about Trump." I only wish I were kidding.

for example. When the Copenhagen conference collapsed in failure in 2009, the press corps' interest in covering global warming collapsed with it. Figuratively fell off a cliff. The *New York Times* actually shut down its standalone environment and energy news section after that event, and that's just one expression of this industry's ongoing attention deficit disorder. This is all well documented. The world's extinction crisis is another example of a far more consequential story worthy of intensive and ongoing coverage, of daily headlines, even. But that's not happening. The political saga that started in the United States in 2016 is a massive ongoing distraction.

"Give the audience what it wants" was something I heard a lot toward the end of my beat reporting career, a clarion call from multiple media corners. This is catastrophically incorrect, short-term thinking. This is how to alienate people and lose their trust and why the audience you do manage to keep eventually loses all respect for you and your institution and starts drifting away. I've promised myself to not repeat these mistakes.

In my experience, to be a competent, respected journalist, you don't give your audience what it wants. That's infotainment, not information. Maintain proper consideration for who your readers or viewers are, of course, but aspire to give them what they *need*, not what they want. That's my advice: give your audience what it needs. And what might that be? Well, I don't know—that's left for serious, curious journalists to discover for themselves. But I definitely know what they don't need: readers can absolutely, positively do without the 11 billionth report about how you-know-who embodies all that is evil and wrong with the world. Give your readers, viewers, and listeners what they *need*, not what they want, and you will go far in journalism—I hope.

And one more minor recommendation, if I may, for you aspiring reporters out there: lay off Twitter. Like, *way* off. Immediately. Moving forward, use it judiciously only if you must, especially for news alerts and press releases. Twitter is great for announcing things that are happening or will happen or for drawing attention to new reports or research recently published and of possible interest to a wider audience. But never, ever again run a report about something some strangers said somewhere on Twitter—you know, the "Twitter blows up about . . ." clickbait. Ninety-eight percent of Twitter's brain farts are written by less than 2 percent of the U.S. adult population—it's not representative of mainstream society. Maintain a healthy distance from social media and you will instantaneously become a better journalist for it, I promise you. It's unfortunate, but in my experience the least informed reporters and editors that I have ever met were the most prolific "tweeters."*

End rant.

Thank you very much for your patience; I appreciate it. That was all very long-winded and terribly cranky, I know, but there is a point embedded there, believe it or not.

Aside from their therapeutic value to me, these ramblings and gripes really do have a place in this book. Yes, the topic of the book is the Endangered Species Act, endangered species, cranes, extinction, and the global biodiversity crisis. Press criticism is very relevant to this discussion, I firmly believe, because how the world views and pursues protections for endangered species—how we choose to face and fight the biodiversity crisis—depends a lot on how the public perceives (or is aware of) the severity of the problems facing the planet's biodiversity today. Public perception matters greatly, and public perception is largely shaped by what the media chooses to—or chooses not to—focus on.

In other words, the attitudes and attention spans of the press corps have enormous influence on what the public pays attention to and what it doesn't (or what it can't or won't pay attention to) and on how politicians and policymakers react. The media also has a very big role to play in setting and maintaining the overarching political mood or atmosphere of a voting public and body politic, the very background reality in which conservation legislation is drafted and implemented in the first place. After all, the U.S. Endangered Species Act was not created in a vacuum, and the state of American and world media in 1973 was far different from what it is in 2023. In other words, I *am* going somewhere with all this ranting, so bear with me for a bit longer, if you please.

Once I was no longer able to keep off the Borg's radar, it quickly dawned on me that it might be best (and wise) for me to begin closing out my beat reporting days and pursue my long-held plans for a career shift. I was and am still grateful to all the folks in the media world who took a chance on me in the first place and let me report for them for so long, but by mid-2019, I had become thoroughly fed up with the industry's one-track mind. So I made plans, and in relatively short order, I extricated myself from that field and latched onto a newer, even better road ahead. I'm now determined to slake my thirst for new knowledge, new insights, and new adventures, even in

* My favorite example is the junior editor and Twitter addict who told me in early 2020 that I shouldn't cover COVID-19 too heavily because it was never going to be a big story. That person confidently said this to me because Twitter said it first. Thus the danger in becoming too hooked on social media—you become less informed, not more so.

subjects and fields beyond my comfort zone. Sure, many other industries or parts of the economy have far too often failed to keep things in perspective these past few years as well, but mainstream U.S. journalism is arguably in worse shape.

Hopefully, things have turned around for the media by the time this book makes print. I'm actually quite hopeful it will, eventually. But looking back, I don't regret any of the time I spent as a reporter. I don't put myself on the same pedestal with the heroic journalists who lost their lives or very nearly did so delivering the stories of the oppressed or distressed to the rest of us, and I never will. But my newswriting career had plenty of interesting up-and-down moments. It was a fun roller coaster ride for sure.

In northeastern Colombia, I was provided a military escort to a contested conflict zone that the Colombian army was occupying at the time, a region that had for a long time been under the firm control of the FARC, the Spanish acronym for the Revolutionary Armed Forces of Colombia. FARC was really just a sophisticated drug trafficking organization masquerading as freedom fighters (pronouncing the acronym as "farce" is appropriate). The stretch of territory along the border with Venezuela that I visited in the rainy and sweltering Catatumbo was still under constant attack by FARC guerrillas at the time, and the soldiers I rode along with even took me to a river crossing to show me what was left of a wrecked bridge the narcos had destroyed with a car bomb two weeks before my visit. I was ushered from location to location in the back seat of an armored pickup truck, sandwiched between two heavily armored soldiers wielding enormous automatic rifles. Three more heavily armed and armored soldiers sat in the truck bed just behind us. Ours was just one in a convoy of about ten vehicles moving from site to site. Whenever we would stop to stretch our legs, sentries jumped out and positioned themselves around our perimeter, scanning the jungle for any signs of movement.

Why on Earth would I deliberately put myself in such a situation? To produce a report on biofuels, of course.

At the time, Colombia's government was encouraging the cultivation of oil palm as feedstock for a national biodiesel program partly inspired by Brazil's massive government-initiated biofuels industry. Colombia also saw biofuels crops as potentially viable and even powerful alternatives to illicit crops like coca, the main ingredient for cocaine. So I traveled to the Catatumbo to interview Colombian farmers who previously grew coca for FARC. When I visited the region in 2011, they were growing oil palm instead for Colombia's nascent but expanding biodiesel industry, so I wanted to see what this all looked like and ask these farmers what they made of this activity. I also had the pleasure of interviewing the general in charge of the Colombian military's Catatumbo division, as well as the nation's minister of defense in Bogota. Of course, the rising popularity of palm oil threatened sizable portions of Colombian rain forest, as well, thereby threatening Colombia's already teetering biodiversity. The authorities I spoke with at the time insisted that oil palm cultivation would be expanded only to lands already under cultivation, ideally replacing illicit crops. I wanted to believe them, but that's what oil palm enthusiasts in Malaysia and Indonesia say as well, and environmentalists very publicly accuse the oil palm industry in those nations of fueling deforestation.

Regardless, I very much enjoyed my time in Colombia. I found the people there to be very intelligent, friendly, and hard working, a nation with an industrious, ambitious population that will one day become an economic force to be reckoned with.

On another foreign reporting trip, I arranged to accompany U.S. Army and Air Force relief flights into northern Pakistan following a series of devastating flash floods that wiped out dozens of bridges in the mountains, leaving several northern communities cut off from the rest of the country. A U.S. Blackhawk helicopter crew, some Pakistani rangers, my interpreter, and I flew into some of the remotest corners of Kohistan to deliver desperately needed supplies to some of these cut-off communities. These relief flights were scheduled to continue for about another month or so, until some of the roads into the area could be reopened and normal commerce resumed.

I was always the first to hop out of the chopper so I could take photographs of all the action. Most of the time locals rushed past me to help the soldiers unload supplies, smiling gratefully but otherwise ignoring me entirely. So I was surprised when during one stop a man with a long red beard and shaved head interrupted me as I snapped another set of relief action shots (I'm actually a terrible photographer but needed some pictures to illustrate my reports, however poorly). He was quite animated, I recall, shouting excitedly and pointing to my left, past a hill, toward the outskirts of town. The noise of the helicopter rotor made it hard to hear, but I wouldn't have understood a thing he had to say anyway because I don't speak his language. So I ushered this gentleman instead to a Pakistani soldier and waited. The two conversed very briefly, and then the soldier's face turned to stone. He looked in the direction that the man pointed. The soldier then gave me a hard stare and gestured to me with his thumb that I had better strap myself back into the Blackhawk helicopter as quickly as possible. I did as I was told. A few more hand signals and some shouting later, and we were all back in the air, without dropping off any of the goods those villagers were waiting for.

I learned what that was all about only after we had returned to a military base just south of Tarbela Dam. Red beard was apparently a Kohistan village elder. He was trying to warn us that a group of armed and aggressive men were lying in wait for us to begin offloading our cargo, aiming to take advantage of that vulnerable moment to likely launch an attack. This village elder apparently objected to this, since it could have resulted in innocents dying in the crossfire while putting an end to deliveries of outside assistance for his community, so this man decided to risk his life for the sake of ours. Since I was useless, he relayed this information to the Pakistani soldier instead, thus our hasty retreat. I don't know if the group lying in wait was Taliban insurgents, some other armed government opposition group, or simply bandits, but I am grateful for that gentleman's intervention. And I honestly don't know what happened afterward, but the Pakistani soldiers assured me that they had every intention of returning to that same village, but next time with better security and only after ensuring that that particular drop site was free of potential threats.

But what was the point of all this? I was dispatched to northern Pakistan by a brilliant editor and former *Wall Street Journal* reporter to put together a story on climate change.

By some estimates, 20 percent of the country had been hit by devastating floods that year, the worst flooding episode in Pakistan's history, and at the time scientists were growing confident that they could tie such a violent meteorological phenomenon to rising global temperatures. The U.S. military agreed—prior to my trip, the Pentagon had issued a report predicting that U.S. troops most likely would find themselves increasingly deployed to weather disaster zones of the sort that had occurred in Pakistan that year. So the editor I wrote for at the time—a man with true imagination, a sense of adventure, and a proper appreciation for what's important (all qualities increasingly rare in journalism)—sent me over there to see what this type of military-led disaster relief work looks like and to hopefully get a glimpse of the U.S. military's future.

The timing wasn't great—that same week, some nut job in Florida was making news by threatening to burn a copy of Islam's holy book in front of any camera-wielding reporter stupid enough to give this insignificant individual free airtime. But that media sideshow didn't matter to my editor; he and I sought to give our readers what they needed, not necessarily what they wanted, so off to Pakistan I went. There I learned that the Pakistanis had more than climate change to blame for the flooding. Decades of uncontrolled, mostly illegal tree felling have left vast swaths of northern Pakistan badly deforested, the hills and mountainsides shaved bare as if by a giant

Norelco. Countless species were likely lost to this process, as well. It also left the downstream cities and towns vulnerable to massive flash flooding. At the time of my visit, it was estimated that up to a third of Pakistan was flooded at some point. This disaster was unprecedented when I visited the country in 2010, but it was later sadly repeated in 2022. In addition to the Army Blackhawk helicopter operation, I also was hosted by the U.S. Air Force on a relief flight from Islamabad to Gilgit. Aside from the hairy landing, that part of the trip wasn't very memorable since we never even left the Gilgit airport, which is a shame because northern Pakistan is beautiful, as are its people. I met some of the most brilliant, warm-hearted folks there during my all-too-brief stay. I'd love to return someday, hopefully seeing far more forest and wildlife than I did on my first trip.

Before those two overseas assignments, there was Haiti. Pakistan's bald mountainsides reminded me of Haiti quite a lot.

Prior to the devastating 2010 earthquake, Haiti was known best for suffering through centuries of poverty, political instability, and ecological devastation. This unfortunate history ultimately resulted in Haiti becoming the first and only nation in the Western Hemisphere to host a United Nations peacekeeping mission. That mission, which went by the French acronym MINUSTAH, had a mandate to quell and disarm violent gangs threatening the peace, to train and build an effective and competent national police force (Haiti had dissolved its military by this time), and to support the government in any other functions deemed necessary. It's no exaggeration that the UN really did run the place during those years. A former boss of mine hated the UN and did his best to avoid covering anything happening at UN headquarters in New York, leaving that all to me, but when I organized his reporting trip to Jacmel in southern Haiti in 2005, he called me up one day to check in and express how surprised and impressed he was. "The UN is king here!" he exclaimed during one of our calls. A few years later I would discover the same thing during my own extended travels in that country.

So why Haiti? Again, the story was environmental in scope.

Whereas other nations fuel themselves with gas and electricity, Haiti powers itself mainly with charcoal, thus the nation's ecological devastation: most of the native forests there have been destroyed, and its wildlife was apparently destroyed with it. A flight over the countryside in a UN helicopter revealed to me row after row of bald hills and mountainsides and the consequent erosion and soil degradation that came with this deforestation and resultant flash floods, repeating disasters becoming commonplace in Haiti as in Pakistan.

Haiti's dependence on charcoal was beginning to have international consequences at the time of my visit. Increasingly, bald spots in the middle of once-forested lands appeared in neighboring Dominican Republic as a thriving and illicit trade in Dominican charcoal took hold, straining relations between the two countries. Shortly before my visit, a Dominican landowner fatally shot three Haitians caught felling his trees and making charcoal in a makeshift kiln they'd built. The local police found some of their bodies charred in that same kiln. Violence became more common along the Haiti–Dominican Republic border, and the UN tried to find ways to put a stop to it. One idea involved a "sustainable charcoal forest" run by the UN Food and Agriculture Organization in Haiti's northeast. Another peacebuilding contingent of MINUSTAH organized a way to recycle paper and cardboard waste into hockey-puck-sized briquettes that could be burned for cooking in lieu of charcoal. Neither project exists now.

Elsewhere, MINUSTAH struggled to stop devastating urban flash floods, which had become routine due to the destruction of surrounding forests. In Gonaives, the peacekeepers hired locals to build water breaks and channels on hillsides surrounding the city in the hope of slowing down future floodwaters. In the city itself, millions in foreign aid dollars were spent on a major infrastructure project, a new floodwater channel to run underneath a main city street. The new concrete tube was supposed to carry flash floods safely under Gonaives, rather than over it, giving the water a route to the sea without taking any unfortunate residents with it. I don't know if that project was ever completed or not. It wouldn't surprise me one bit to learn that construction was ultimately called off entirely. The recycled paper briquette project was turned over to the Haitian government when MINUSTAH eventually left, and the initiative was abandoned shortly afterward. Reforestation efforts in Haiti have come and gone. Mostly gone, I'm afraid. There have even been efforts to promote ecotourism in some remaining pristine spots of the country, most under the sea, but these efforts were dealt setbacks in the form of deadly earthquakes and political strife.

So why am I wasting your time with all these tales? To illustrate something important, something very important to the main theme of this book, in fact.

This book you now hold is about species and biodiversity, the abundance and variety of life on Earth, and human efforts to save other species from destruction, namely from extinctions caused by us. And in all my travels overseas, throughout my years of real reporting, I almost never encountered any wildlife during my time spent outside in the field. Anywhere. No single majestic nonhuman living thing that could linger on in my memories long after I returned home from Haiti, Pakistan, or Colombia. Not even a squirrel.

The only exception to this pattern was in Kenya, where a colleague and I were sent to report on happenings in a southern Kenya national park famous for abundant biodiversity: Amboseli National Park. There I saw elephants, lions, zebras, giraffes, and more. I saw other amazing animals during a tour of the much smaller Nairobi National Park—rhinos, baboons, and hyenas even. But this was an exception because the Kenyan authorities make a point of protecting these lands and their wildlife from hunting and other human disturbances. Another key lesson of this text is the importance of policy, policy decisions, and carefully maintained habitat protections.

Everywhere else I went? Nothing. I never encountered a single other living creature that wasn't human during my explorations of various rural corners of the Caribbean, South America, and South Asia. I walked through remote corners of the Colombia–Venezuela border region and can't recall ever coming across even birds, let alone some crawling critter like a tapir, capybara, or other native fauna. I was eaten alive by plenty of nasty and potent mosquitos, however, and still have marks on my ankles to show for it.

I flew high over vast portions of wild spaces in northern Pakistan, areas far removed from civilization, and never saw so much as a single deer. Drive through the Rocky Mountains in Colorado, and you'll witness all sorts of beautiful and majestic creatures—bighorn sheep, elk, moose, and if you're really lucky an occasional bear or mountain lion. Not so much in the rugged mountains of Pakistan, apparently, which closely resemble the Rocky Mountains at higher reaches where the tree fellers had yet to arrive.

The same was true during both of my trips to Haiti. At remote corners of the border with the Dominican Republic? Nope. Flying over land in a UN helicopter to Gonaives? Nothing. In the fields surrounding the small rural village of Thomonde? Sorry. I saw nothing in or around that village, either, except goats and chickens, all domesticated livestock, naturally. I even stopped along the route to Thomonde to tramp through the bush for a bit so I could take a picture of a dam, an old World Bank–sponsored project built to power Port-au-Prince. I didn't notice anything then, either, not even the chirp of a sparrow. This was all rather eerie for me, having grown up in the U.S. Mountain West. Well, I should mention that, outside Kenya, I've never encountered examples of wildlife from my reporting trips conducted *on dry land*.

My first Haiti journey in 2009 involved a scuba diving excursion off the coast of the island of La Gonave, a bit northwest of Port-au-Prince. The point of this investigation was to witness firsthand the relative health of

Haiti's coral reefs and to determine whether this part of the country might one day see a revitalization of the tourism industry (the country's north is already home to a popular privately owned cruise ship stopover). So during my first visit to the country, a local dive master, my driver, a few friends, and I drove beyond Port-au-Prince, past a former Club Med that then served as a weekend retreat for Brazilian peacekeepers, to a point near the community of Arcahaie. From there, we set off on a boat for La Gonave. Three of us later donned scuba gear and jumped into the water. Only two of our dive team knew what to expect. I quickly caught on.

The underwater world of Haiti is beautiful, for sure. The coral reefs are not in the best shape, at least not where we dove, but they are not completely lost, either. But what really struck me was the relative lack of fish. There were fish, of course, but mostly smaller specimens, few and far between and nothing like the bounty of aquatic species I encountered at Saba, the Dutch Caribbean Island where I earned my open-water scuba diving certification. Elsewhere in the Caribbean, divers routinely encounter sharks, rays, groupers, parrotfish, and more. Those species were absent entirely off the coast of La Gonave when I investigated.

I do recall seeing a single lobster; the abandoned, busted lobster trap next to it suggested that these waters once hosted a far greater abundance of these crustaceans than at the time of my visit. The biggest aquatic specimen we encountered was a lonely lionfish looking for something to eat. Many of you reading this now understand that this was a bad sign. That lionfish species is native to the Pacific and in fact is considered a devastating invasive species in the Western Hemisphere, wreaking havoc on other reef communities throughout the Caribbean, Gulf of Mexico, and, increasingly, the waters of South America, including Colombia's coastline and reefs. The lionfish invasion has gotten so bad that every year U.S. government officials temporarily waive a ban on fishing in the Flower Garden Banks National Marine Sanctuary off Texas (run by the U.S. National Oceanic and Atmospheric Administration, or NOAA) for the sole purpose of allowing spear fishers to kill as many lionfish as possible within a week. They call it the Lionfish Invitational. I was actually looking forward to joining one of these trips as the first journalist to ride along and capture lionfish alongside NOAA-approved divers, but I had to depart Texas before experiencing this opportunity.

These species weren't the examples of biodiversity that I was hoping to encounter in Haitian waters, but it's what we found: a few tiny fish, one lonely surviving small lobster, and an invasive lionfish covered in poisonous barbs, a species no fisherman wanted to have anything to do with at the time. The coral reefs were decidedly lacking in larger native fish species, thus it's safe to assume that this corner of Haiti will not become a hub for scuba diving tourism anytime soon. What this area needs most is government control and protection over the species that remain, how few and far between they may be. The same goes for anywhere else you and I may travel where we find plenty of people but barely any wildlife to speak of.

That I haven't encountered even traces of wild animals in most of my past reporting trips serves to illustrate the severity of our current biodiversity crisis.

Though we are accustomed to reading depressing conservation news, the reality of the situation is far worse, considering that it's possible to visit a remote corner of the extremely rainy Catatumbo in northeast Colombia and not even remember hearing birds in the trees or to fly over the most inaccessible parts of mountainous northern Pakistan without seeing any four-legged animals taking advantage of the aquamarine rivers draining the meltwaters of the nearby Himalayan glaciers. As mentioned, Haiti once had a Club Med and an active scuba diving and snorkeling scene. In fact, didn't Bill and Hillary Clinton spend their honeymoon there? I think that's the case. You can still scuba dive in Haiti, of course, and even get certified there, but you won't find the kind of rich underwater biodiversity that you would encounter only a few hundred miles away on other Caribbean islands. That's because the Haitian government—to the extent that it exists—simply cannot or will not do what it takes to ensure sustainability in these coastal regions. I have a sense that Haiti's economy will rebound when its ecology starts to do so—that's one lesson I took from my back-to-back trips there (the second time I arrived by hitching a ride on a UN plane flying out of Santo Domingo following the terrible January 2010 quake, where I failed to encounter any biodiversity during my travels in and around Thomonde).

We are amid the sixth great mass extinction event. It's correctly characterized as a crisis. Laws and policies like the Endangered Species Act help us turn a corner on this crisis in large parts of the world. The whooping crane and the red-crowned crane are still alive and doing much better than they were in the past thanks to laws and policy choices. But most of our planet remains hostile territory to endangered plants and animals. Even in the United States, once ubiquitous species like oak trees are deemed increasingly threatened by human-driven ecological and climatic changes.

Are we too late? I don't believe so. I certainly hope not. But first, it's important that we comprehend the scale of the problem, a disaster so massive that I easily stumbled upon it during my various travels without intending to.

Just how bad is the global biodiversity crisis? The word "crisis" probably doesn't do it justice. By all accounts, it's a rolling catastrophe. And the

American press corps doesn't spend nearly enough time drawing attention to it, in my opinion. Even today, you-know-who dominates the headlines, while recently published and troubling reports about the worsening condition of the planet's biodiversity go largely unnoticed.

Extinction is natural and inevitable. Our kind will also become extinct one day—that's right, we will all someday end up tossed back into the sea, reduced to tiny motes and molecules, forgotten shapes never to be seen again. Most traces of our civilization will become buried and lost within 1,000 or so years following our exit from reality. Yes, extinction is inevitable, but most of the time, extinction is imperceptibly slow, so slow that evolution, which is slow and imperceptible in its own right, usually outpaces extinction to win the race: the tortoise beating the slower tortoise. That's why the fossil record generally shows past biodiversity increasing over time, at least until a major extinction event occurs. But extinction is ever present.

As evolution and biodiversity expansion press forward, extinction comes along for the ride; except in more natural circumstances, species generally aren't driven to extinction at rates that threaten overall ecosystem health and function. But that's no longer true. In fact, there's broad scientific consensus that the current rate of species extinction that we are witnessing today is at least 1,000 to 10,000 times faster than any natural background level, or at a rate that could be expected if only nature was the driving force and not humans.

A current rate of species loss running 1,000 to 10,000 times faster than normal, natural background rates—that's the recent estimate put forth by the IUCN. They should know; they compile and monitor the famous Red List of global plant and animal species declared threatened or endangered.

According to an IUCN 2017 brief, at least 16,900 known species of animals and plants today face the prospect of annihilation, possibly ceasing to exist entirely within our lifetime. This is likely a gross underestimate because IUCN simply doesn't have a complete accounting of all species in existence—not even close. In some places, assessments simply aren't allowed—in North Korea, for example, where a starving population quite possibly has already eaten every variety of fish and fowl within reach. But we do know enough to understand with increasing confidence that you and I now find ourselves in the midst of a sixth mass extinction. Future paleontologists may ultimately name it the Anthropocene mass extinction event, because there is no doubt that we now exist in a human-dominated epoch and that humans are the driving force behind the alarming rate of species losses now being witnessed across the globe.

Here are some rather depressing facts contained in that dated yet still very relevant IUCN report, one of those reports that goes ignored or unnoticed by the mainstream press corps:

- We know for a fact that humans have driven at least 870 species into oblivion during the past 500 years. Many of these were iconic animals like the great auk, Tasmanian tiger, and Caribbean monk seal. This figure will probably hit 900 to 1,000 species soon. Hopefully the vaquita won't join this list, but I'm afraid it will. Many species that we don't yet even know exist almost certainly will, as well. For the IUCN and others, the most worrying aspect of our current extinction crisis is the thousands of extinctions that might occur without us ever becoming aware of them.
- In just ten years, IUCN added more than 3,200 species to its Red List of endangered and threatened species.
- One in four known mammal species are at risk of extinction.
- One in six known bird species are at risk of extinction. Many species of birds that were endemic to specific islands have already been lost, including, most famously, the dodo, but also several species of birds once endemic only to the Hawaiian Islands.
- One-third of all known species of amphibians are facing probable extinction.
- Though it's fashionable to blame climate change for everything these days, there is overwhelming evidence that current rapid rates of extinction are being driven first and foremost by habitat loss—humans converting previously wild lands into farms and urban spaces. Though climate change is a massive and exacerbating threat, habitat loss is overwhelmingly to blame, not global warming. My students get this one wrong constantly, another failure of the news media perhaps.
- It must be said that global biodiversity faces pressure from multiple fronts, all human caused. Habitat loss is the biggest driver for certain. But the world is still plagued by overhunting and overfishing, wildlife trafficking, pollution, alien species threatening native flora and fauna (like the lionfish), and, yes, global warming.
- The extinction crisis is being felt worst in places like Brazil, still-developing China, India, and Southeast Asia. Some academics are fond of blaming the Global North for the problem, but these days the crisis is felt most acutely in the Global South. What these places have in common are very large human populations, high rates of economic development and land-use changes, and previously impressive levels

of biodiversity now under constant assault by economic growth. They also have very weak species protection laws or governments that overwhelmingly prioritize development over all other considerations. You can't blame colonialism for all this, though a finger of blame can be pointed at capitalism and Western economists' obsession with growth over everything else.

- There are very strong correlations between human population growth, rapid rates of development, and species loss. And in this case, correlation *is* causation.[2]

So things are really bad out there for sure. Much of the above goes a long way toward explaining why I never happened upon any wild animals during any of my treks through rural Colombia, Haiti, or Pakistan, while encountering abundant if occasionally mismanaged biodiversity at Kenya's protected public parks.

That's where this conversation turns more hopeful—appreciating the scale of the disaster while also appreciating how some extinctions were averted entirely and what we can learn from these case studies.

CHAPTER TWO

~

The Crisis and the Mystery

Can extinctions be prevented? Of course, if just temporarily (since we must be honest here, eventually all life will be rendered extinct, including human life). For a long time, they can be halted and the relentless march toward extinction can even be reversed. The International Union for Conservation of Nature (IUCN) lists several examples that it considers among the more promising global species preservation success stories.

As I highlighted earlier, humans today care a great deal about whether other animals are driven from the face of Earth because of us. Thanks to this evolved (pun intended) attitude and carefully crafted conservation initiatives like the Endangered Species Act, we still have the European white-tailed eagle, a species of kestrel native to Mauritius, the Hawaiian goose, and the crested ibis of China. Other success stories mentioned by IUCN include the white rhinoceros and the Indian vulture. The Mekong River in Southeast Asia is home to a species of catfish so massive that it can grow to weigh more than three times my weight, and I'm not a small guy. Humans almost wiped out these massive catfish completely, but thanks to other humans we can still travel to Southeast Asia to see this monster of a catfish should we choose to do so.

Humans also prevented the extinction of the whooping crane of central North America and the Japanese red-crowned crane of eastern Hokkaido, as I mentioned previously. You'll be reading a lot about these two beautiful species in the pages that follow.

What's the key behind all these success stories? Well, it's rather messy and complicated, as I indicated earlier.

"Globally threatened species frequently require a combination of conservation responses to save them," as IUCN explains it. "These responses encompass research, species-specific actions, site and habitat-based interventions, policy responses, and communication education." Above all, we need better habitat protections, since it's primarily habitat loss that drives most of these species to peril. As IUCN puts it, "it is much more effective and economical to protect habitat in the first place than to try to restore it after it has been destroyed or to reintroduce a species."[1] This view seems reasonable.

The IUCN concluded a World Conservation Congress in Marseille, France, an event that was delayed by a year due to the COVID-19 pandemic. Aside from electing new leadership and handling other administrative matters, at this congress, IUCN announced the latest updates to its famous Red List. Which means the information I included here has already changed and will likely change again by the time this book is published. But as of 2021, the IUCN Red List of threatened and endangered species now includes 138,374 species, according to an IUCN press release, of which 38,543 are threatened with extinction, or more than a quarter of the living creatures listed on the Red List.

Some tuna species were upgraded, including the Pacific bluefin, a heavily overfished species nearly annihilated by the global sushi craze—95 percent of the bluefin tuna's original biomass has been lost, but new stock assessments and political pressure now has IUCN classifying this species as near threatened rather than vulnerable. That's not where I would put this species on the Red List spectrum given that the total biomass of bluefin tuna was once 95 percent larger, but I'm no expert here. Several species of sharks and rays are being hit hard, IUCN acknowledges, and we may lose a few of them entirely.

Other species are faring better, but on balance, things are getting worse.

For example, the Komodo dragon of Indonesia is now considered an endangered species. Prior to 2021, it was classified only as vulnerable. Predictably, IUCN blamed climate change but also acknowledged that rapid development adjacent to government conservation areas is shrinking the size of the habitat considered vital to the ongoing survival of the world's largest lizard.

Habitat loss, biodiversity loss, rapid rates of extinctions—this is an ongoing global problem, so it requires a global solution, right? Not necessarily.

"Global problems require global solutions" is a favorite catchphrase at the United Nations, but logically it doesn't follow. Why *must* this be the case? Why can't global problems be resolved by a set of local solutions? In fact, waiting for the world to unite as one in order to address some massive problem far too often makes things much worse in the interim rather than better. I should know—I've seen this occur firsthand in the halls of the United Nations, in the debate chambers where draft agreements are put through the ringer well before they reach the floor of the General Assembly. Do global problems really require global solutions, or is there a better way? This question will come up again later in this exploration.

From its point of view, IUCN very much believes that the biodiversity crisis requires a global solution, one in which the whole world works in unison toward one precise goal. How hard could that be, right? To press this point home, delegates at the 2021 gathering in Marseille drafted, adopted, and then released to the world "The Marseille Manifesto" upon concluding their talks.

Merely six pages long, the Marseille Manifesto is replete with the same popular catchphrases, platitudes, and seemingly profound yet ultimately meaningless jargon frequently encountered in such documents. That should be expected (trust me). But dig a bit deeper, and you will find some meat hidden among the dry bones.

To their credit, the delegates have come up with concrete proposals that for decades have been lacking in prior gatherings of governments party to the Convention on Biological Diversity. For instance, the manifesto notes the hundreds of billions of dollars that governments are spending to prop up economies hit hard by the COVID-19 pandemic and asks that at least 10 percent of these funds be earmarked for domestic conservation initiatives; namely, to purchase, set aside, and then protect and properly manage habitat critical to species' long-term survival. This is an excellent idea.

The manifesto's drafters also want to see at a minimum 30 percent of the world's surface—all lands, lakes, and oceans—set aside in some protective manner by the year 2030, which isn't that far away, mind you (this goal was later codified at the 15th Conference of Parties to the Convention on Biological Diversity at the end of 2022 in the Kunming-Montreal Global Biodiversity Framework agreement). Achieving this feat means that governments must agree to establish the world's first marine protected areas in portions of the oceans beyond Antarctica, essentially designating patches of open international waters as entirely off-limits to fishing and resource extraction of any sort. There actually exists a great body of scientific evidence showing that establishing and enforcing these types of fishing fleet no-go zones does wonders for recuperating aquatic biodiversity and restoring ocean reef health. The manifesto also envisions a complete and eternal ban on deep sea mining in international waters, a move that would effectively shut down the International Seabed Authority, a UN organization based in Kingston, Jamaica.* Whether you agree with any of these ideas or not, they are all concrete, precise proposals that even a cynic like me can sink his teeth into.[2]

Will the world come together to do what IUCN says is required to halt biodiversity loss by 2030? I've been monitoring the United Nations since 2004. My experience includes seven years following international diplomacy at the UN from its headquarters in Manhattan, New York City. I've spent hours—hell, days and even weeks—listening to diplomats negotiating decisions and drafting outcomes. I once found myself in the basement of the UN until 3:00 a.m. on a Saturday morning waiting to see if delegates would agree to put in place a moratorium on bottom trawling, a particularly destructive fishing practice that causes near-permanent damage to the ocean floor (imagine mowing down a forest that won't grow back for 100 years with a tractor in an effort to capture ground squirrels and you'll have a sense of what deep ocean bottom trawling entails). A moratorium on bottom trawling fishing seemed like a sensible policy to implement at the time, but the diplomats I was busy interviewing ultimately rejected the idea. And I know for a fact that governments party to the Convention on Biological Diversity have been largely ignoring the CBD for most of this treaty's existence—the UN itself admits as much, even if CBD signatories do not.

Although I may find this new IUCN manifesto interesting and somewhat promising in terms of whether nations will change course in response, given my experiences, I'm sorry to say that I'm not holding my breath. But there are some promising indicators.

France, host nation to this particular IUCN Congress, says that it will set aside 30 percent of its territory for conservation and species protections by 2027, including 5 percent of that portion of the Mediterranean Sea under Paris' control.[3] We can easily follow up on these promises to see if they've actually been kept and later hold to account the government leaders who made the promises in the first place if they aren't. Issuing promises that other people have to keep is far too common in international diplomacy, so I find France's recent declarations to be rather refreshing.

Yes, the world's biodiversity continues to shrink. The situation is getting worse. Rates of extinction remain far too high. But people are aware and committing themselves to doing something about all this. It's a start.

---

* I won a National Press Club award for my reporting on the International Seabed Authority prior to writing any portion of this book. Seabed mining is the latest potential threat to deep sea species and another worthy story that the mainstream news outlets are mostly ignoring.

Meanwhile, let's not continue holding out hope for the arrival of a miraculous global solution to this global problem. Let's explore local solutions instead. There we may be surprised to find far more traction and progress.

Before researching this book, I was most familiar with United Nations declarations and treaties, especially global environmental treaties like the Basel Convention and the Paris Agreement. They may have some things in common, but generally speaking, international environmental treaties are nothing like domestic environmental ordinances.

With domestic democratic politics, governments are free to do what they want so long as they can get a majority of lawmakers to align themselves in a certain direction or to support certain decisions. If opposition voices find themselves in the minority position, well then, they're out of luck—their voices will be heard, but their ideas or objectives will be omitted entirely from the final document, so long as a majority of standing representatives agree to an action of some sort.

This usually isn't the case at the United Nations.

There, member governments try to achieve everything by consensus, and getting more than 190 governments to unanimously concede to anything isn't an easy trick. So it can take a very, very long time to get anything done at the UN.

For example, UN member states have been debating for close to 15 years now whether or not to establish any specially designated marine protected areas in international waters as a way of controlling overfishing, to somewhat turn the tide against the ongoing plundering of our oceans. More than 25 years ago, the International Seabed Authority was told to draft rules for mining the open ocean floor; it still hasn't as of this writing, though it may ultimately succeed in doing so by the time this book reaches your hands.

When UN environmental agreements are sorted out in much shorter time spans, the results usually leave everyone disappointed. Oftentimes, diplomats simply come up with "something" deemed adequate enough after failing time and time again to meet greater political and public expectations through international diplomacy. They'll still hoot and shout and applaud their historic "successes" in front of the TV cameras, as we all saw back in 2015 when the Paris Agreement on climate change was adopted, but the actual outcome is much less impressive upon closer inspection.

Still, every now and then UN member states can surprise us and agree to documents and outcomes that are fairly comprehensive and impactful. The

United Nations Convention on the Law of the Sea (UNCLOS) is an excellent example.

UNCLOS took about a decade to negotiate, but the resulting document gives clear instructions for how nations may lay claim to ocean territory, marine exclusive economic zones, and extended subsea continental shelf resources, in addition to spelling out just how maritime disputes are to be addressed bilaterally or arbitrated at the International Tribunal for the Law of the Sea, headquartered in Hamburg, Germany.

But more often than not, agreements like UNCLOS are the exception to the rule. UN environmental agreements are usually cumbersome and seemingly weighty yet ultimately empty documents that leave much to be desired by observers like me who know that, when it comes to environmental protection, clear policy with precise guidance is where the rubber hits the road. UN environmental declarations and treaties are too often the opposite of this. They can be lofty in language and decidedly lacking in any real specificity—rich in vision and ambition, while utterly devoid of any usable details for the actual practitioners of conservation. Or to put it another way, all fluff and no substance. The 1992 Convention on Biological Diversity is probably a good example of this trend.

Much of the language in that landmark environmental treaty can be interpreted in multiple ways. For instance, article 11 of the CBD states that: "Each Contracting Party shall, as far as possible and as appropriate, adopt economically and socially sound measures that act as incentives for the conservation and sustainable use of components of biological diversity." The CBD's signatory states are largely left to interpret this and other clauses like it as they see fit, and to be fair, given the vagueness of the language, that's probably their only recourse. It's because of wording and diplomatese like this that the CBD stands out among global treaties for its near-universal participation (sans United States, which has signed but not ratified this multilateral agreement) and very low levels of actual compliance.

The CBD does direct its signatories to set aside protected areas to help conserve species, but it offers zero guidance about how they should go about doing this or how much territory they should set aside, and it only asks them to do so voluntarily "as far as possible and as appropriate," which is clever get-out-of-jail-free language used over and over again in the treaty that essentially permits governments to ignore clauses entirely if they like. The CBD declares that resources found within a nation's territory belong to that nation and can be exploited by that nation however that nation may wish. The treaty asks only that countries exploit their resources sustainably, or "at a rate that does not lead to the long-term decline of biological diversity, thereby maintaining its potential to meet the needs and aspirations of present and future generations." It calls for signatories to develop "national strategies, plans, or programs" for preserving biodiversity. Few of them actually have.

The extinction crisis rolls on and ecological biodiversity nearly everywhere still suffers long-term decline, so the vast majority of member governments appear to be ignoring the CBD. This could be because the CBD doesn't include any credible monitoring or compliance provisions anywhere in the language. This was very deliberate. The Convention on Biological Diversity is mainly an aspirational treaty replete with aspirational language. UN environmental treaties are severely limited in scope and function in ways that I explain later in this book. This is a consequence of the realities of international relations and the system of global governance that exists in our world in lieu of a one-world government (no "Leviathan" to enforce its will on the planet). The CBD is not international law per se, but rather is best described as a document that encourages states to conserve ecosystems and biodiversity while facilitating information sharing among interested governments—nothing more, nothing less.

By comparison, the 1973 Endangered Species Act (ESA) of the United States is a finely crafted precision instrument. You could set your watch to it.

For sure, the ESA no longer exists in its original form. For instance, it was amended in the 1980s so that some members of Congress could direct a portion of federal endangered species protection funds to be spent not on the actual protection of endangered species, but rather on stocking rivers and streams with fish to the eternal delight of sportfishing enthusiasts such as myself. But the hard, precise provisions remain.

The ESA empowers the Department of the Interior (specifically the Secretary of the Interior), through the U.S. Fish & Wildlife Service, to list species as either endangered or threatened with extinction. It then directs Interior to identify listed species' critical habitat and then to draft a plan for preserving both the habitat and the listed species. The draft plan must be presented to the public prior to adoption for public comment. The law clearly spells out precisely how much time the Department of the Interior has to achieve all this.

The ESA also details how the Department of the Interior is to accept outside petitions for either listing or delisting species, essentially inviting the public to hold the department's feet to the fire. Notices about listing decisions must be published in the Federal Register and even in the local newspapers circulating in areas where the species is deemed to be present. The federal Endangered Species Act also spells out how federal officials are to cooperate with state officials on listing decisions and recovery plans, and it voids all state laws that conflict with or are in contravention to the Endangered Species Act. However, it leaves state governments free to establish their own species preservation laws, so long as they don't appear to dilute what federal authorities are trying to accomplish—in other words, state governments can enact and enforce laws that are tougher and even more restrictive than the Endangered Species Act, but not less so, according to the act itself.

The ESA is widely acknowledged—by those minimally knowledgeable enough to appreciate it—as a classic example of a well-rounded and thought-out piece of domestic legislation, obviously drafted by folks who had a clear sense of the outcomes they were trying to achieve while staying mindful of preexisting legal and political precedents or limitations. For instance, it's a common misperception that the ESA places too heavy a burden on private citizens, landowners, and state governments. In reality, the law is laser-focused on how federal authorities must proceed in the event of an endangered species listing. This can result in regulations that could, for instance, restrict how close federally permitted oil and gas drilling activity can occur on or near formally declared critical habitat, but in such instances the oil and gas industry also can intervene to make its case known well before any such regulations are implemented, and the government must hear these arguments and objections per the ESA. Conservative lawmakers and states' rights advocates may gripe and bitch about the onerous nature of the Endangered Species Act and may even dream of gutting it, but they have yet to provide even one bit of convincing evidence to show how the act restricts freedom or impedes economic development to any extent.

Let me repeat this because it's an important point: there is little to no evidence that the U.S. Endangered Species Act has kept anyone from making money or prevented any proposed private commercial development from proceeding. I challenge anyone reading this now to prove me wrong, because I've looked and have yet to come across any concrete, convincing examples of this happening. This is a rather important point worth highlighting: the current sour political mood and the preponderance for ideologically driven hyperbole egged on by an ideological press corps could make it impossible for the United States to achieve anything like the 1973 breakthrough ESA today (unless lawmakers managed to pass the Recovering America's Wildlife Act after I wrote this). In fact, the social-media-driven ideological radioactive cloud that comprises the largest portion of atmosphere enveloping the United States in our times is perhaps the greatest threat to the Endangered Species Act's future. Let's hope it blows over quickly.

Less well known, the 1973 Endangered Species Act is remarkably international in its scope, outward looking and even diplomatic to a degree, which also could never be accomplished in the halls of the U.S. Congress today.

The ESA requires the Department of the Interior to inform foreign governments of endangered species listing decisions if a species is also present in that foreign government's territory. Foreign governments are even invited to participate in the subsequent public comment period, that time frame whereby interested outsiders can make their opinions known to the government officials proposing a new rule or regulation.

I know of no other instance in any other country in which foreign citizens and foreign government representatives are invited to the table when another nation's domestic environmental laws and regulations are up for consideration. One big exception might be when governments endeavor to earn UNESCO World Heritage status for certain protected sites and thus invite outside scientific scrutiny, but even this would be an example of multilateral United Nations oversight that's entirely voluntary, not compulsory. The ESA legally mandates that foreigners be invited to participate in the listing process. The next time one of your overseas friends talks about how America acts too unilaterally and cuts itself off from the rest of the world, show that person the Endangered Species Act and then ask if his or her own government might one day invite American citizens to participate in the drafting of its domestic environmental rules. I could be mistaken, but I doubt there are many similar examples of this.

The ESA explicitly names some of the countries that the United States must continue cooperating with on species protections per earlier signed and ratified agreements (Canada, Mexico, and Japan are the three nations named in the ESA). The text declares that the ESA itself is the primary instrument through which the U.S. government complies and cooperates with the Convention on International Trade in Endangered Species of Wild Fauna and Flora, or CITES, arguably the strongest multilateral environmental treaty in existence. It gets even more interesting. The ESA says that the Secretary of the Interior is free to use foreign currency held in U.S. accounts to pay for endangered species conservation programs happening in foreign countries (the Department of the Interior can tap U.S. dollar funds for this, too, but must use foreign currency accounts first). The law even encourages state governments to cooperate with foreign governments on species conservation programs, and it offers federal money to help facilitate this in an attempt "to develop and maintain conservation programs which meet national and international standards," as the text declares.

The Convention on Biological Diversity is international in scope, obviously, but so is the U.S. Endangered Species Act. As a domestic piece of legislation crafted at a time when American politics, public discourse, and the press operated quite differently from what we see today, the ESA stands out as a shining example of how government can be harnessed to solve even the most vexing problems facing our planet, with the right approach. Indeed, the Endangered Species Act can be used as a lens to better understand what is wrong with the current state of international environmental diplomacy and why massive environmental challenges like climate change or the global extinction crisis never seem to get resolved. Language-wise, both the U.S. Endangered Species Act and the UN Convention on Biological Diversity could use some work—let's face it, one was drafted by lawyers, the other by diplomats. But compared to the Convention on Biological Diversity, the U.S. Endangered Species Act is a precision-crafted, finely tuned instrument. The CBD is too often a rhetorical mess. The ESA is like a vehicle crafted by engineers; the UN's CBD, more like a vehicle literally designed by committee. The ESA has been proven to work to date. The CBD? Not so much, unfortunately—and by the UN's own admission, to that world body's credit.

Yet the 1973 Endangered Species Act, now celebrating its 50th birthday, is not a perfect piece of legislation, either in form or implementation. Far from it.

The U.S. Fish & Wildlife Service admits that ESA-backed efforts to save some species from extinction have failed on multiple occasions. Listed species have disappeared from the planet despite an Endangered Species Act listing. Far more have been rescued from oblivion, but just barely. One example I'll delve into later is the Attwater's prairie chicken. It's a species closely related to the lesser prairie chicken, except the Attwater's prairie chicken finds its home in the hot and humid coastal prairies along the Gulf of Mexico, whereas the lesser prairie chicken prefers drier habitat farther inland. The last time I checked in on the Attwater's prairie chicken, it was possible to take every single living specimen in existence on Earth and fit them all into the back of a cargo van. The species is mostly kept going by a busy captive breeding program in cooperation with places like the Houston Zoo. It was very nearly rendered extinct in the wild by Hurricane Harvey in 2017. This endangered species rehabilitation program is an example of a federally funded conservation initiative that's been spinning its wheels for decades, likely to the consternation of lawmakers who have no doubt spent tens of millions of dollars trying to prevent the Attwater's prairie chicken's extinction. Things may finally be turning the corner for this species today, though it's too soon to tell.

And then there are examples of nations succeeding above and beyond the achievements of the Fish & Wildlife Service, without the benefit of any ESA-type law. It may be possible to learn from these examples as we try to decipher the lessons learned from 50 years of the Endangered Species Act and how we humans might best proceed to put the ongoing sixth mass extinction (the Anthropocene mass extinction event) to bed by 2030, as the Marseille Manifesto tells us we must.

For example, in a quiet corner of eastern Hokkaido, far off the beaten path of the hustle and bustle of modern Japan, there exists a species of crane: the Japanese red-crowned crane. It is all but identical to the famous whooping crane of North America. The main difference between these two species is that the red-crowned crane tolerates the cold better and sports darker plumage on some parts of its body. Other than this, it is hard to tell the species apart. My own students in Japan have a hard time telling a whooping crane from a red-crowned crane apart.

The red-crowned crane and whooping crane also share remarkably similar stories, or conservation histories.

Both species were nearly driven to extinction by the early 1900s. By the end of World War II, individuals of both species numbered no more than in the dozens each, if that. In fact, the Japanese had long assumed that the red-crowned crane had vanished from their islands completely until a government-led expedition in Hokkaido stumbled upon a small flock one day. Both the red-crowned crane and whooping crane have been subjects of active, well-funded, and concerted efforts to prevent their extinctions and drive their population recoveries, efforts led by relatively wealthy governments, with these initiatives each lasting for more than 70 years on opposite ends of the planet and supported by competent conservation scientists. And both of these conservation efforts have been successes—neither species is facing the prospect of extinction today.

However, one of these programs has seen far more success than the other. Why? This is a question worth exploring in some detail.

Seventy years on, what do these two endangered species success stories look like today?

As I type this, the U.S. Fish & Wildlife Service and Canadian authorities can celebrate their milestone of reaching more than 500 individual whooping cranes in the central migratory flock, the largest concentration of whooping cranes in North America. It's a milestone that anyone involved in whooping crane conservation and rehabilitation can be rightly proud of.

They should feel proud—believe me, I have the data that proves they've achieved something remarkable.

Japan, meanwhile, is looking at somewhere between 1,800 to 2,000 red-crowned cranes living in Hokkaido today, along with tentative signs that the red-crowned cranes are poised to begin migrating beyond the only Japanese home they've known for more than 100 years. Red-crowned cranes may begin nesting on other islands of the Japanese archipelago before this book publishes. Japan achieved the greatest momentum in this species rehabilitation story in the absence of an equivalent to the Endangered Species Act; meaning since well before the 1970s Japan's signature endangered species has been rebounding in population size far more robustly than the whooping crane, a signature ESA-listed and -protected species.

Today, the whooping crane is still classified as an endangered species by both the U.S. Fish & Wildlife Service and the IUCN. The Japanese authorities upgraded the status of their beloved red-crowned crane sometime in late 2020, removing it from the official "endangered" status. The IUCN reportedly followed suit in mid-2021, upgrading the red-crowned crane's status in its famous Red List. Red-crowned crane populations are assessed as declining on the Asian mainland but increasing and rebounding on Hokkaido. Several decades after beginning nearly in tandem, these two conservation stories diverged sharply. While the whooping crane of North America is still technically an endangered species, its cousin in northern Japan is no longer so classified.

American (and later Canadian) and Japanese conservationists have been kept busy for some seven decades managing two parallel endangered species recovery projects on opposite ends of Earth. Their stories closely align, almost eerily. These two populations of near-identical crane species were almost rendered extinct by the early 1900s. In the 1950s, each population barely numbered in the dozens of individual birds. And beginning in the 1950s, both conservation efforts became far better organized and systematic, launching in earnest at about the same time. Although whooping crane numbers and Hokkaido red-crowned crane numbers were initially comparable at the start of these organized recovery programs, somehow the Japanese officials have managed to grow their red-crowned crane population to almost quadruple the size of the present-day American and Canadian whooping crane migratory flock. On the American side, the whooping crane is still endangered, including the largest central migratory flock. On Hokkaido, the picture is quite different today.

How, or why, did this happen? That's what I'm trying to figure out. In fact, it's a rather important question that we all should consider more deeply and carefully, because answering this question may reveal for us some key clues about how to stop the catastrophic biodiversity crisis rolling still.

Let's dive a bit more into these two conservation stories. They're both success stories. But one has been rather more successful than the other.

CHAPTER THREE

The Antisocial Crane

In late 2021, the U.S. Fish & Wildlife Service (FWS) formally proposed that the ivory-billed woodpecker be declared extinct, meaning gone for good. The International Union for Conservation of Nature, or IUCN, had yet to make up its mind on this question by the time of this writing (IUCN still had the species listed on its Red List as critically endangered, meaning basically on the verge of extinction). But I think it is fair to say that FWS has it correct here, sadly.

This once splendid species endemic to the southeastern United States and Cuba hasn't been sighted in America since the 1940s. In fact, there's a high probability that this species was already long gone by the time the U.S. government got around to listing it as endangered, so it wouldn't be fair to call this extinction an example of a failure of the Endangered Species Act per se. A final sighting in Cuba reportedly occurred in the 1980s.

The ivory-billed woodpecker wasn't just another bird. This species was one of the largest species of woodpeckers in the world. It was beautiful. Now it no longer exists. The usual culprits—human stupidity and callousness, otherwise known as pointless overhunting and habitat destruction—sealed the ivory-billed woodpecker's fate, long ago perhaps, robbing you and me of any opportunity to enjoy this marvel of evolution that once was.

How many times does this have to happen? How many more species will be added to that list of the annihilated? I don't know, but I do know that the extinction list will have greatly expanded well before this book reaches the market and your hands. Yes, more amazing creatures will join the ivory-billed woodpecker's fate.

Still, conservationists have saved from the brink several large species just as amazing and iconic as this one. Perhaps the most famous example is the peregrine falcon, believed to be the fastest animal in the world, capable of dive-bombing at prey at velocities of more than 300 kilometers per hour. Humans almost destroyed the fastest animal in the world through heavy use of the chemical pesticide DDT in agriculture, but the species has rebounded strongly and now the IUCN lists the peregrine falcon as of "least concern" on its famous Red List. The peregrine falcon is an ESA success story, I believe, given that most recovery efforts happened under ESA jurisdiction.

North America's largest bird was also almost driven to extinction. Its closest cousin in Japan very nearly shared the same fate.

The whooping crane (its name translates to "American white crane" from Japanese) is the largest bird in North America in terms of standing height, though not wingspan. The whooping crane is big and elegant and remarkably tall, with adults roughly matching the height of an average-sized adult human.

East Hokkaido's red-crowned crane is the same—big and beautiful and a sight to behold, especially when they're flying just a few meters or so over your head. These birds were almost lost to time for good, as well, before the humans inhabiting Japan got their act together. A separate population of red-crowned cranes makes its home in eastern Siberia and northern China, but most experts agree that population is in decline, the continued victim of hunting, poisoning, and habitat loss. On Hokkaido, the situation is markedly different from that of the Asian mainland. Red-crowned cranes in Japan have since recovered nicely.

Work to conserve both of these crane species—whooping cranes and red-crowned cranes—isn't over yet, though in Japan they are now beginning to have that discussion. In America, there's no question that conservation will continue for the foreseeable future even as FWS mulls removing whooping cranes from the endangered species list. The whooping crane population hasn't bounced back by anything close to the degree seen with the peregrine falcon, though the Fish & Wildlife Service has done a remarkable job at saving the whoopers from extinction, and the species' future prognosis is looking very good.

Fate first brought me to both these birds.

After seven years of reporting from New York, I relocated to Houston, Texas, in late 2011 to focus more on energy stories, in particular the oil and gas industry. The shale oil boom had begun in earnest, especially in the Eagle

Ford, and the latest Permian Basin oil boom was only just getting started. Offshore energy exploration was also booming along nicely, at least until oil prices crashed a few years after my arrival to the Lone Star State. But oil and natural gas were not my only interests. My curiosity in conservation policy never went away entirely. Naturally, I looked around for a case study that would allow me to delve a bit more deeply into the practical implications of certain environmental policies and conservation practices. Texas is home to dozens of these case studies, perhaps even thousands. The story of the whooping crane was a convenient one, given the proximity of the Aransas National Wildlife Refuge to my then home. Thus I was drawn to the whooping crane for the first time.

Later I would follow my wife back to her home in Japan and to East Hokkaido. There, you cannot avoid encountering the red-crowned crane and the Japanese people's strong attachment to this magnificent species. Being the ever-curious journalist I was, I couldn't help but notice not only the similarities in size, appearance, and behavior of these species of crane, but also their remarkably similar conservation and recovery histories. This is how I initially became drawn to the story of the Japanese red-crowned crane—it was, in a way, a repeat of the whooping crane story, only happening an ocean away.

My encounters with both these avian species were fortuitous, I think. A side-by-side analysis of these cousin species, the whooping crane and red-crowned crane, serves as a very useful alternative to any experiment in endangered species management in illustrating what might happen when certain conservation practices are implemented and others are not.

To run this sort of experiment properly, you would divide the same endangered species population into two groups, one serving as the control group and the other the experimental group. Next, to determine the effect of a certain management practice, you would use the management technique on the experimental group over a period of—oh, say—50 years or so, while denying the control group the benefits of that same management practice. At the end of this experiment, you would look at what happened to the two separate groups and tally your results. Is the experimental population larger than the control group? Smaller? Healthier? More threatened than ever? In an ideal situation, this is how one might run this type of scientific inquiry.

In reality, actually doing such an experiment with an endangered species is probably illegal, and few would have the patience to run it for half a century anyway. But the more I delved into the stories of these cranes, the more I came to realize that, in a sense, the Fish & Wildlife Service and Japan's Ministry of the Environment pretty much ran this very experiment for me, beginning long before I was even born, without knowing it. So I made it my mission to get to know as much about these two remarkable birds and their stories as possible.

What follows is a wonky, part-narrative account of what I've found so far. And I also have to mention that the following overview and all the research it's based on, including some of the more technical details included here and in other parts of this book, first appeared in a paper published in the *Journal of International Wildlife Law & Policy*.[1] I later delivered a presentation of this paper and my research at the Wildlife Society's 29th Annual Conference in Spokane, Washington, in early November 2022. For any curious academics reading this now, I invite you all to read that paper about my initial research, if you're curious enough and would like an even wonkier take on the investigation that ultimately inspired this book.

The famous Lone Star State flag. Vast open spaces and plenty of blue sky. Insane drivers flying down the freeways way too fast. Truck stops and barbecue joints. Snowbirds from Minnesota cruising in their RVs looking to escape the cold, snowy Midwest winters. These are some of the more common sights one finds while driving throughout Texas. Here's another one: "Keep Out" signs.

Texas is America's second-largest state in both area (after Alaska) and population (after California). There is much to see and enjoy there. In truth, I find Texas to be highly underrated in terms of its natural beauty—the Lone Star State doesn't receive nearly enough attention in that regard. But the state does suffer from some serious limitations.

For starters, though it's vast in size and natural diversity, Texas has only one true large natural lake, Caddo Lake, of which it shares half with Louisiana. There are other smaller natural lakes, which some may consider to be just ponds (Powderhorn Lake is a good example). Caddo Lake is incredibly beautiful and highly recommended, though I would advise against camping there in August, a mistake my wife and I made once. It's a beautiful lake but the only large natural one in the state—all other major Texas lakes are man-made reservoirs (the Dallas metro area famously built scores of reservoirs to hedge against droughts, to get ahead of a rapidly rising population, and to enrich a handful of politically connected contractors, most likely). That it's lacking in natural lakes is just one downside to living in Texas, though the state makes up for it with the Gulf of Mexico. But there's an even bigger downside: Texas also contains within its vast expanse the lowest percentage of publicly accessible land of any of the 50 states. I may be technically incorrect here, but Texas must certainly be near the top of that list.

More than 95 percent of Texas is privately owned and generally off-limits to the public. That leaves just 5 percent or less of this massive state's territory for the citizenry to play on, not counting the waters of the Gulf of Mexico. Texas is home to two amazing national parks, Big Bend and the Guadalupe Mountains, but they're far from any of the state's main population centers. Big Bend National Park is about a ten-hour drive from Houston. I've briefly visited the Guadalupe Mountains, but only because my wife and I happened to be driving through the area. Texas state parks are generally closer, and they are also immensely popular. Some of my favorites include Brazos Bend, Colorado Bend, and Enchanted Rock. The state's federal wildlife refuges are also a bit closer to where people live and certainly are worth a visit.

That's what initially brought me to the Aransas refuge—curiosity about the whooping crane coupled with a desire to stretch my legs on a vast expanse of natural, publicly accessible land, of what little there is that Texas has to offer. Perhaps you've also seen it or will get a chance to someday soon.

The Aransas National Wildlife Refuge is a little gem of public land. You'll find it stretched along the state's southeastern Gulf of Mexico coastline in the region between Victoria and Corpus Christi. Because the vast majority of Texas is fenced in and marked with "private property: do not enter" signs, the state's residents flock in great numbers to the few state parks, wildlife refuges, and other public lands accessible to them on weekends, and Aransas is no exception. There are picnic areas, hiking trails, and a small interpretive center run by the Fish & Wildlife Service. Visitors can enjoy viewing the refuge's coastal marshes from the bird stands set up along the water's edge. Aransas is not a bad place to spend a day or even just an afternoon, and the state's residents are certainly aware of this. It's for this reason that we chose to make our journey in the middle of the workweek, to avoid the worst of the crowds and traffic.

Aransas encompasses approximately 46,300 hectares (or nearly 115,000 acres) of coastal wetlands, estuaries, grasslands, and salt marshes along a relatively undisturbed stretch of the Gulf of Mexico. It's home to a wide array of wildlife and plant species, as is the area more generally. And there are plenty of other attractions to enjoy outside the refuge. For instance, there are some neat places around Corpus Christi, including Mustang Island State Park. My wife and I camped on the beach at Mustang Island once, the first and only time I've ever camped on a beach. Very early the next day we enjoyed watching federal wildlife conservation authorities and their partners in the nonprofit sector release scores of tiny endangered Kemp's ridley sea turtle hatchlings in the Gulf's waters, carefully keeping the baby sea turtles safe from human gawkers and seagulls. Not all of the baby sea turtles would make it, for sure, but the U.S. government is spending big in the hopes that enough of them will survive and thrive long enough to help expand the Kemp's ridley sea turtle population, perhaps to the point where the species will someday no longer require federal interventions of any sort.

Though it's huge, the Aransas National Wildlife Refuge is not contiguous. Rather, it's divided into four subsections in the same general geographical region.

The main section of Aransas encompasses the entirety of a large, fat peninsula extending into the surrounding bays. A small lake is found on this peninsula, as well, and farms run right up to the border of the refuge. Another much smaller section of the refuge can be found on the Lamar Peninsula to the west, also home to Goose Island State Park and "the Big Tree," a massive ancient oak tree kept partly aloft with the help of support beams. Visitors to the area should check out Goose Island State Park and the giant oak tree, too, if they have the chance. Matagorda Island is a coastal barrier island separating San Antonio Bay from the Gulf of Mexico, and it's home to the second-largest section of the refuge. There's no telling how many hurricanes Matagorda Island has survived. Rounding out the official boundaries of the Aransas National Wildlife Refuge is a tiny subsection located apart and alone, farther to the east, at a little corner of protected land adjacent to the ghost town of Indianola and abutting Powderhorn Lake. And Powderhorn Lake is the namesake of Powderhorn Ranch, the newest addition to the whooping crane's protected habitat, which I'll get to in a moment.

As mentioned earlier, the Aransas National Wildlife Refuge is home to a wide variety of species, but this corner of Texas is most famous for being the main wintering grounds of the only wild migrating flock of whooping cranes left in North America. For the purpose of this book and for my academic research, I refer to this whooping crane flock as the "Aransas-Wood Buffalo" or AWB population. Others do, as well. I also occasionally refer to these birds in both the singular and plural tenses as "whooper" or "whoopers." Two other separate flocks of whooping cranes can be found, one in Louisiana and another that, with human assistance, made its way between Wisconsin and Florida, but as far as I'm aware, that effort lost funding and the human-led Florida-to-Wisconsin whooping crane migration no longer occurs. That group is now divided into a managed Wisconsin flock and the small remaining population left to its own devices in Florida. The AWB flock is the only population of whooping cranes that resides in Canada for part of the year. My research mainly concerns how they spent their time in the United States under the Fish & Wildlife Service's protection. Studying whooping cranes in Canada is more difficult, because they spend their summers in remote, hard-to-reach areas of Wood Buffalo National Park.

AWB whooping cranes arrive at the Aransas National Wildlife Refuge every year in winter, usually beginning around December, to forage at the refuge and to escape the bitter cold of central Canada. While wintering in Texas, whooping cranes mainly stick to the coastal parts of the refuge to dine on the blue crabs that crawl out of the mudflats for feeding time when the tide rolls out. But AWB whoopers venture back and forth farther inland to forage wherever it pleases them. They spend quite a bit of time feeding on the neighboring farms, for instance. AWB whooping cranes also eat the fruit of the wolfberry plants scattered throughout the refuge. They are omnivorous waterbirds, so they eat almost anything they can get their beaks on: frogs, fruits, bugs, fish even.

Once spring rolls in, the AWB whooping cranes depart their winter vacation spot at Aransas, taking several weeks to migrate throughout central North America—parts of Kansas, Nebraska, the Dakotas, Montana, and occasionally venturing further east—until landing in the northernmost section of Wood Buffalo National Park, Alberta/Northern Territories, Canada.

Wood Buffalo is North America's largest national park, mostly situated in Alberta but with a smaller section cutting into the Northwest Territories. That's where the AWB whooping crane's breeding primarily occurs. As of this writing, I have not made it to that remote corner of Wood Buffalo National Park to see in person what the whooping crane's territory looks like there—they mostly spend their time in the northern stretches of the park and at spots just outside the park's boundaries—but from what I've read and seen online, their habitat at Wood Buffalo is not dissimilar from the wetland way stations where whooping cranes rest while migrating through the Midwest twice a year.

Though this is not the only flock of whooping cranes in North America, AWB is by far the largest and most important; it's considered the central whooping crane flock, and it happens to be a focus of my research. The conservation effort surrounding the AWB whooping crane also has the longest history to speak of, and it's centered at the greatest amount of protected land set aside specifically to prevent this particular species from extinction, once considered inevitable.

As mentioned, the Aransas National Wildlife Refuge is huge; it's one of the largest Department of the Interior–managed wildlife refuges in the United States. Wood Buffalo National Park is even more massive; it's roughly the size of Connecticut. And although it's technically not the largest terrestrial national park in the world (that honor goes to a massive chunk of Greenland), it is certainly up there. As noted above, Wood Buffalo is North America's largest national park. Keep all of this in mind as you read this book—the AWB flock enjoys a huge swath of territory of ideal whooping crane habitat, and the Endangered Species Act is essentially founded on the well-established scientific principle that protecting and maintaining critical habitat is crucial to the survival and recovery of most critically endangered species. Protecting and managing critical habitat is central to what the U.S. Fish & Wildlife Service does.

Again, throughout this read I reference the AWB flock frequently, to refer to this specific flock of whooping cranes resident in central North America, which migrates twice annually from Canada to Texas and back. But I also occasionally use the shorthand "whooper" or "whoopers" to refer to the same thing. Because, why not? Just for kicks. So please keep this in mind as well; I don't want you thinking that I've somehow become obsessed with a sandwich from Burger King halfway into this read. Just a heads-up.

Now on to a bit of history.

⁓

The Aransas National Wildlife Refuge was established by presidential executive order in 1937, during the thick of the Great Depression. It was developed partly by the Civilian Conservation Corps (CCC), which was set up through a U.S. federal government program designed to provide paying work to folks struggling to find employment anywhere else. In other words, the CCC was basically a Depression-era New Deal program aimed at getting struggling, out-of-work single men using their muscles to help build facilities and infrastructure for conserving America's natural heritage. CCC laborers built hiking trails, access roads, park facilities, rest stops, and more. It was a good idea that the government may yet be forced to repeat someday considering how unstable America's economy has become in recent decades.

Initially established with the aim of protecting a variety of species of migratory and nonmigratory waterfowl, the Aransas refuge was quickly identified as key to the survival of the critically endangered whooping crane very early in its creation. Today, the Aransas National Wildlife Refuge is known to exist primarily to keep whoopers alive, even though these animals spend only about three or four months of the year at the refuge proper. U.S. Fish & Wildlife maintains the refuge for other important and valuable species, but for eight or nine months of the year, their job is to maintain the refuge for the AWB cranes even when no whooping cranes are present, all in preparation for their annual return in winter.

As I already mentioned, the whooping crane is the largest bird in North America, but only in terms of height. The California condor, another ESA federally protected endangered species, has a larger wingspan. Like the California condor, whooping cranes were nearly wiped off the face of the earth by the usual suspects that I point to over and over again in this book: uncontrolled, unfettered, mostly pointless hunting and habitat loss. Habitat loss mainly resulted from Midwest farmers draining or filling in thousands of wetlands along the crane's flyway in order to grow crops. The hunting was even less excusable: whooping cranes are big, slow, and easy to shoot, and I don't recall ever reading rave reviews of whooping crane meat in any culinary article. Hunting them could hardly be called sport. Throughout the 1800s, mindless, careless yahoos shot whooping cranes with abandon, not necessary for food or for money, but because they were lazy, never truly respected nature, and figured that they could get away with it. And for the most part, they did. Some still hold these attitudes. As I was writing this chapter, four AWB cranes were reportedly shot dead in Oklahoma. If the man (always a man) responsible for this illegal act of poaching is ever caught, chances are good he'll face only a slap on the wrist, as with past instances.

By the mid-1940s, the AWB whooping crane population was said to have fallen to as few as 16 individuals. Just 16 remaining in the entire world. This put the species on the brink of extinction—depending on the population's ratio of males to females or adults to juveniles, by the end of World War II, the AWB whooping cranes either could have slowly recovered, possibly with too little genetic diversity in the population, or could have faded into oblivion regardless of any measures federal officials might have taken to save the species. It could have gone either way, but the federal government thankfully decided to take a stand and a chance on the whooping crane, to at least make an effort to save the largest species of bird in North America from complete annihilation. Federal conservation efforts began in earnest around this same time, especially from 1950 onward.

Initially, AWB whooping crane conservation and population recovery proceeded in fits and starts. Federal wildlife authorities basically had to learn as they went, and a lot of what they tried during the earliest days of the whooping crane recovery effort was experimental, methods largely never tried before. AWB whooping crane conservation got off to a very slow start. In 1952, there were still only an estimated 21 AWB whooping cranes in existence on Earth, according to FWS data. Their numbers had risen only slightly from what they were back in the 1940s, but at least the population seemed to be heading in the right direction by then, increasing and not steadily in decline.

Fish & Wildlife attempted a variety of interventions to help save the whooping cranes.

Whoopers typically lay two eggs. Both hatch, but generally only one of the fledglings survives and reaches adulthood to help reproduce another generation. It's as if one baby whooping crane is born only as a sacrificial lamb to whatever limiting factors or other vagaries of nature exist or control whooping crane population growth. Predators, disease, or perhaps a combination gets them, or they're the runts of the litter, unable to compete with their siblings for nutrition.

Understanding this, conservationists and federal officials at Aransas quickly learned that they could take one of the two eggs from a nest without irreparably impacting population growth. In fact, they realized that doing so might even help increase whooper numbers by hatching those eggs in an incubator and then breeding the newborn chicks in protective captivity before releasing them back to the wild as healthy adults, ready to fend for themselves. Early in the AWB flock's conservation history, that's exactly what Fish & Wildlife did, at least for a while. They also briefly dabbled with artificial feeding of the species, understanding during the program's early days that easier food availability could enhance an animal's chances of survival in the wild. But the artificial feeding experiments quickly came and went in the 1960s; Aransas managers ultimately abandoned that effort entirely, and it's never been repeated, or at least that's what they've told me.

Americans generally don't like the idea of feeding wildlife. We've been taught from a young age that it's distasteful and bad for the animals. Federal and state conservation authorities continue to promote this idea; that's why you frequently encounter signage instructing park visitors to keep their distance from animals and never, ever to feed the wildlife, under any circumstances. The message has been repeated so often that it's become nearly second nature to most of us. Though it still happens on occasion for other species on other federally protected lands, artificial feeding is generally frowned upon in federal wildlife management in the United States, as there's fear that members of a species may become too habituated to people or too dependent on artificial feed for their survival. It's also seen as something that you just shouldn't do in general. Animals in nature must learn to survive off nature, after all. Right?

During my travels in Kenya, I saw visitors at Nairobi National Park gleefully feeding a pack of wild baboons. I was appalled, shocked even, that this concept of minimal interference in the survival of wild animals wasn't as thoroughly ingrained into the minds of citizens of other nations as it is in my homeland. And Kenya in particular has been hailed for its conservation practices. The concept of never feeding wildlife is deeply ingrained in North American environmental and public consciousness. Maybe this view is outdated or too rigid? Maybe exceptions can and should be made? Who knows?

At any rate, by 1952 only 21 whooping cranes called the Aransas National Wildlife Refuge their winter home, according to the U.S. Fish & Wildlife Service's records. The AWB whooping crane population steadily started trending upward from this time on in annually recorded head counts.

In 1966, the U.S. Congress passed the Endangered Species Preservation Act, and the whooping crane was among the first species to be listed for protection under this act in 1967 (the ivory-billed woodpecker also was listed that same year). When the whooping crane was first officially listed as an endangered species per federal law, the AWB population numbered only 48 cranes, according to FWS data. By this time, the initial tentative experiments with direct artificial feeding of whoopers at Aransas began and then quickly and quietly ended. As far as I know, they only tried it for a few weeks of the year then stopped. The captive breeding program for AWB cranes would later end, as well, though captive breeding in support of the separate Louisiana flock of whooping cranes would later be established and continued until at least 2017 (perhaps it resumed later after I wrote this).

I categorize artificial feeding and captive breeding as direct population management approaches, meaning these are management techniques through which conservationists try to directly manipulate an endangered species population's survival and/or rate of reproduction.

At Aransas, direct population intervention was largely set aside as the Fish & Wildlife Service shifted primarily to a strategy of habitat management. I define habitat management as interventions that largely forego efforts to directly manipulate individual survival and population growth in favor of interventions designed to enhance the habitat and natural environment's capacity to support endangered species' chances of survival and reproduction. It's a more indirect strategy. Thus, by shifting primarily to a strategy of habitat management, FWS determined that its job was to focus mainly on helping nature help the whooping crane survive and recover over time, however long that might take.

That is, rather than providing food for the flock or feeding fledglings in captivity until fully grown, FWS officials shifted course early on. For decades now, they've done their best to improve survival conditions at the Aransas National Wildlife Refuge to the point at which nature presumably would provide the birds with all the winter sustenance they could ever require.

The AWB whooping crane's listing as a specially protected endangered species was carried forward as the 1966 Endangered Species Preservation Act officially gave way to the tougher and more comprehensive federal Endangered Species Act of 1973, a landmark piece of legislation at the time.

To this day, the whooping crane is considered a symbol of the Endangered Species Act. If you go to the Wikipedia page for the Endangered Species Act, the first photograph you'll encounter is that of a whooping crane (at least that was the case when I wrote this sentence). The AWB whooping crane population has steadily risen since federal protections were enhanced under the newer, tougher, and more detailed Endangered Species Act. Canadian wildlife authorities didn't get around to designating the whooping crane as an endangered species according to their own national conservation laws until 2000.[2]

Throughout the history of AWB whooping crane conservation, habitat management has been the main tool used by the Fish & Wildlife Service to ensure this crane's survival. However, it's important to clarify here that habitat management has not been the only method employed or experimented with over the subsequent decades of whooper rehabilitation.

As noted, a limited captive breeding program was used in an attempt to boost the number of individual birds in the AWB flock during the early years of whooping crane conservation. Then in the 1960s, as mentioned earlier, the FWS briefly experimented with artificial feeding as well, only to quickly abandon that experiment.[3] But for the majority of the refuge's history, the prime conservation method has been habitat management.

The habitat management methods practiced in defense of the whooping crane are varied. For instance, Aransas National Wildlife Refuge managers occasionally undertake prescribed burning, manual brush removal, and other such interventions in an effort to encourage the growth and propagation of vegetation better suited for this large lanky bird. Whooping cranes can't take off in flight without a running start, so they need a "runway," and woody plants are a major hindrance for them. Or the birds could accidently injure their wings on trees or bushes. Woody brush also provides good places for predators to hide, so whooping cranes generally avoid thick forest cover and prefer open expanses with good lines of sight.

Habitat management, enhancement, and conservation are the tools of the trade at Aransas. It's been that way for a while, and this continues to be the case.

A variety of other such interventions are also taken. For instance, the Fish & Wildlife Service occasionally plants wolfberry plants at the Aransas National Wildlife Refuge to improve the volume of forage available for whoopers during their winter layover in the reserve. In a sense, this is also artificial feeding, but whoopers still have to find and forage at these plants for themselves. The wolfberry plants aren't crops either—they're planted and then left to survive on their own. Droughts can and do get them. There are other habitat management techniques undertaken that aren't mentioned here. Still, some direct population management practices haven't been abandoned entirely. FWS also monitors the birds' health, occasionally checking their vitals, taking blood samples, tagging and monitoring them, and so forth.

One thing is lacking. What federal officials haven't done at Aransas is take measures to manage or otherwise manipulate the hydrology of the coastal wetlands that the whooping cranes call their winter home. This despite the fact that AWB whooping cranes rely mainly on the protein provided by the blue crabs found in the mudflats of Aransas, and blue crab abundance is determined by the balance between seawater and coastal freshwater inflows that help to regulate the salinity of the brackish coastal parts of the refuge's mudflats. Regional hydrology is critical to this ecosystem's health. And yet, as far as I can tell, the federal government hasn't tried to regulate freshwater inflows into San Antonio Bay. It's mostly out of their hands, unfortunately; that's largely the purview of the State of Texas, and Texas authorities don't like it when federal government officials try to tell them what to do with the state's water.

This is a fair point to emphasize. The AWB whooping crane's survival isn't entirely up to the Fish & Wildlife Service. State conservation managers have their role to play, as well, especially in Texas. This reality has led to conflicts in the recent past. I describe one such episode briefly below.

In the 2010s, a coalition of Texas coastal conservation and whooping crane enthusiasts calling themselves the Aransas Project, or TAP, wielded the Endangered Species Act to pursue a lawsuit against the Texas Commission on Environmental Quality, otherwise known as TCEQ.

TCEQ is a state agency based in Austin that's supposed to protect Texas's environment. But that same agency is too often perceived as acting aggressively in favor of private landowners and business interests instead, to the detriment of conservation and environmental quality and in direct contravention of the agency's name and mandate. I personally have no opinion of

TCEQ or hold any views about the agency, but this is an accurate description of its general reputation among the public in Texas and within the greater U.S. conservation community. Whereas other state conservation bodies often lobby in favor of endangered species listings, TCEQ is more famous for its reputation for vigorously attacking federal endangered species listing proposals and other conservation efforts as soon as they're floated.

For example, around the same time the whooping crane lawsuit was working its way through federal court appeals processes, TCEQ went to bat for the state's oil and gas industry to lobby against a federal endangered species listing for the dunes sagebrush lizard. This is a rare species of lizard that exists in only a small pocket of shinnery oak bush and sand dune habitat in eastern New Mexico and western Texas, land that just so happens to sit right on top of the Permian Basin, the most productive oilfield in the United States. The listing proposal happened amid the horizontal drilling and hydraulic fracturing boom that saw the once-written-off Permian Basin roar back to life. Because of the dunes sagebrush lizard listing proposal, oil companies grew fearful that drilling restrictions were imminent, so they and their allies at TCEQ pressed hard against the proposal. TCEQ would be damned if environmental quality for the dunes sagebrush lizard would be protected.

The Fish & Wildlife Service later backed down, as is too often the case—it fears angering congressmen in Washington, D.C., and often doesn't see the point of pressing a matter. In exchange, several oil and gas companies promised to set up a special organization that they would fund. This unique, privately established Texas conservation body then supposedly would work with the industry to preserve the dunes sagebrush lizard. It was a sort of private-sector-led conservation initiative, except led by for-profit oil interests instead of dedicated nonprofits like the Nature Conservancy. This initiative was called the Texas Conservation Plan and was celebrated at the time as the first of its kind and a major breakthrough in state, federal, and private-sector conservation cooperation. It didn't last long. In fact, it never really got started. This short-lived Texas Conservation Plan was mainly designed by the Texas state comptroller's office. Why? I haven't the foggiest clue. There is no earthly reason to expect a state comptroller would be qualified to lead an endangered species recovery plan, let alone design one. But that's what happened. Like I said, a first-of-its-kind effort.

I once visited this now-extinct organization's office in Midland, Texas, and interviewed its sole employee, whose name I sadly can't remember. He was a nice enough guy, likeable, and he happily answered all my questions, if evasively and in circles, that I recall. But he was there mainly for public relations purposes and to provide a thin veneer of legitimacy to the entire operation.

Fish & Wildlife Service officials should have been more suspicious and skeptical of TCEQ and the comptroller's intentions from the get-go, because the Texas authorities quickly deemed their dunes sagebrush lizard conservation plan confidential information. They basically classified it as top secret, not for anyone's eyes. I'm not making that up—they established, launched, and lauded a private-sector-led effort to save the dunes sagebrush lizard from extinction, replete with press releases and other fanfare, and then wouldn't even allow FWS officials—or anyone else for that matter—to actually review their supposedly amazing plan. Here's another lesson from the 50-year history of the federal Endangered Species Act: it exists because state governments can't be trusted to do the right thing.

I now know why the comptroller classified the dunes sagebrush lizard recovery plan as top secret, a risk to national security should that information ever leak. We all do. The State of Texas wouldn't allow scrutiny of its Texas Conservation Plan because this plan basically never actually existed. There was no real plan, as it turned out. It was all so much kabuki theater, and that guy I met in Midland could have been a hired actor for all I know. If so, he deserves an Oscar, I can tell you that.

By 2018, the Texas Conservation Plan was considered by nonprofit interests to be effectively nonexistent. That little office I visited in Midland did jack squat to protect dunes sagebrush lizard habitat. This solved one mystery for me: the reason why that nice man I spoke with looked so bored and why he couldn't point me to any concrete examples of successful conservation interventions undertaken by his organization. He simply had nothing to do, aside from fooling reporters such as myself, and he couldn't point out a single significant measure taken to prevent the lizard's habitat from becoming lost entirely and thus to prevent the species itself from being pushed to extinction. The feds eventually called out the industry and TCEQ on all this, and in 2018 the Texas comptroller's office announced it was "withdrawing" the plan and pulling out of the agreement entirely. The Texas Conservation Plan to save the dunes sagebrush lizard and the organization that supposedly led it were shut down.

TCEQ was deeply involved in this subterfuge, so its reputation in conservation circles isn't all that great, although I'm sure it organizes some wonderful conservation initiatives on occasion.

Anyway, back to the tale of the whooper and its recovery.

As mentioned, the Aransas Project sued TCEQ because the otherwise expanding AWB whooping crane population plummeted during the winter of 2008 to 2009, losing several individuals while Texas experienced a bad drought. More whoopers were lost in 2011 during one of Texas's worst single-year droughts in history, a dry spell that got so bad that some communities in western Texas nearly ran out of drinking water for their residents and were forced to resort to emergency measures.

I was flown out from New York to Texas during the summer of 2011 to cover this episode, visiting that state for the first time to report on the record drought and the conditions it left much of the landscape and communities in. My travels took me to San Angelo and then Llano, where I witnessed the town's reservoir at an astonishingly low level, nearly bone dry. I did my best to conserve water during my one-night hotel stay there but was later dismayed to see a local scofflaw watering his plants, head lowered with a shameful look on his face, as I walked past him to try to find a place to eat.

The Aransas Project (TAP) alleged that TCEQ did nothing during the droughts to ensure that the Aransas National Wildlife Refuge received enough freshwater inflow from the Guadalupe River into San Antonio Bay, as state authorities prioritized the interests of rice farmers drawing water farther upriver. TCEQ countered that it wasn't its job to protect the whooping cranes—it regulates surface water access for farmers and communities, not for federally protected lands. TCEQ basically argued that it wasn't its business if the cranes lived or died. TAP argued that the Endangered Species Act made it their business—that, as federal law, it's all of our business. To be fair, TCEQ could have been technically correct here, but that agency also has a legal obligation to help preserve Texas's natural heritage, whether it's found on state, federal, or private lands. Whooping cranes are part of that Texas natural heritage, so TCEQ certainly could have found a way to allow its surface water management work to continue without any negative repercussions to the whooping cranes. But then again, you'll recall how TCEQ responded to the threat posed to the dunes sagebrush lizard—rushing to the defense of industry, not to the lizard or the environment.

Freshwater river inflow is important to the Aransas ecosystem because it helps regulate or otherwise moderate the salinity of the bay, which in turn affects the salinity of the brackish waters and mudflats that sustain populations of blue crabs. And, of course, blue crabs are the main source of winter protein for whooping cranes. When conditions are too salty, blue crab numbers plummet. This greatly lowers the volume of protein available for the AWB whooping cranes who migrate to this corner of southeast Texas every winter primarily for the purpose of dining on as many blue crabs as they can get their beaks on.

The theory driving the lawsuit was that the drought and lack of sufficient freshwater inflow led to a collapse in the blue crab population at Aransas, resulting in several poor whooping cranes starving to death. TCEQ fought back hard against the TAP lawsuit in court and even roped in allies from the Guadalupe River's state water authority to help in their counterattack. TCEQ's legal argument was essentially as follows: to hell with the whooping crane, that's the federal government's concern and not our problem.

The U.S. Fish & Wildlife Service says that droughts in Texas are the primary causes of the largest losses to whooping crane numbers, more than any other single event or factor in the conservation effort's entire history. Indeed, independent studies support the view that drought conditions can negatively impact the AWB whooping crane population to a significant degree.[4] When river and stream levels are low, the influence of ocean tides on Aransas' coastal hydrology increases.[5] Reduced freshwater inflow into San Antonio Bay is known to result in higher salinity of the brackish waters, which are the main breeding grounds for blue crabs. When these brackish zones become too salty, blue crab numbers fall, and there is far less food for whooping cranes during their winter layover as a result. The science is clear on this question.

The ecology of the Aransas National Wildlife Refuge is heavily influenced by the fine balance between freshwater inflows and coastal tidal saltwater intrusions, which determine the seasonal abundance of blue crabs. Yes, the whooping cranes dine on wolfberries and other forage, but folks I've spoken with pressed over and over again the importance of blue crabs to the AWB population's winter survival. A smaller blue crab population results in less winter protein that AWB whooping cranes require to restore energy reserves following their long migration from Canada. Occasional periods of low rainfall and lower river levels combined can lead to much of the coastal marshes drying out completely, negatively impacting the development of sufficient forage for the whooping cranes, particularly in conditions of extreme drought as Texas has experienced in the recent past.[6]

It was this very sensitivity of the Aransas National Wildlife Refuge to drought that resulted in TAP filing its lawsuit against TCEQ, though from what I've read, the jury was still out on the extent to which Texas droughts hit whooping crane numbers at the time the case was filed. But the facts are not in dispute: during a drought that occurred in 2008 through 2009, the Fish & Wildlife Service estimates that the AWB whooping crane population at the refuge fell by about 21.4 percent.[7] That's a significant blow to this conservation effort. FWS also recorded a steep drop to the refuge's winter whooping crane population in the really nasty 2011–2012 drought, the one I was sent from New York to report about. In particular, 2011 saw the harshest one-year drought to hit Texas in about a century, and that dry spell has been blamed for reducing AWB whooping crane numbers from an estimated 279 birds at that time to some 245 birds, a decline of about 12 percent. There was no direct evidence of a link between Texas droughts, freshwater inflow disruptions, and rising whooping crane mortality at the time, but plenty of circumstantial evidence existed to make a convincing case. So TAP sued.

In *TAP v. Shaw*, plaintiffs called for TCEQ to revoke permits it issued to the Guadalupe-Blanco River Authority, accusing GBRA of failing to ensure sufficient inflow of fresh water to the Aransas National Wildlife Refuge during the drought and thus causing the excessive single mortality event witnessed in the AWB whooping crane flock—initially in 2008 and 2009, but again in 2011, as that event happened while TAP's lawsuit was still working its way through the courts. Though TCEQ was listed as the primary defendant, GBRA intervened in support of TCEQ. This court fight dragged on for years, as such disputes are prone to do in America's glacial-speed judicial system. But TAP's lawsuit ultimately failed and Texas authorities prevailed. The whooping crane survived regardless.

TAP failed to convince the courts to sanction TCEQ and the river authority over the fate of whoopers to droughts. In June 2014, a federal circuit court judge ruled in favor of TCEQ, overturning an earlier court judgment that had been awarded in favor of the Aransas Project. TAP later vowed to take it all the way to the U.S. Supreme Court, while TCEQ vowed in turn to continue spending Texas taxpayers' money fighting the lawsuit, rather than sitting at the table with TAP to see if something could be worked out. TCEQ and its state-paid lawyers prevailed; the Supreme Court ultimately declined to take up the case, thus killing the lawsuit entirely.

So TAP lost in court and appeals and failed to convince the U.S. Supreme Court to review the case, ending the suit entirely by 2015. But that wasn't the end of this long saga.

For some reason, the two sides ultimately decided to bury the hatchet—well, not TCEQ. TAP and GBRA have since agreed to jointly research links between changes in freshwater inflows into San Antonio Bay and conditions that may prove beneficial or harmful to whooping cranes wintering at the Aransas National Wildlife Refuge.[8] Someone at GBRA must have realized that the health of the whooping crane could be rather important to southeast Texas's economy. As far as I know, these studies are ongoing and even may have been given a shot to the arm in the form of cash from the 2010 Gulf of Mexico oil spill settlement between BP and the federal government. One certainly hopes this initiative yields some fruit and doesn't experience the same fate as that earlier and mostly fake dunes sagebrush lizard conservation initiative.

Although state authorities ultimately have a role to play, whether TCEQ agrees or not, unfortunately FWS often has to operate without state cooperation, as this episode demonstrates. And they've been succeeding despite this handicap. After TAP's lawsuit was tossed from the courts, FWS pressed on with its habitat management strategies anyway, and the AWB flock's numbers continued expanding despite the setbacks dealt by the Texas drought and TCEQ's lack of interest in helping to preserve this magnificent species.

A side note, if I may.

Interestingly enough, though it's been shown that the Aransas National Wildlife Refuge is particularly susceptible to harm from droughts, Aransas has proven remarkably resilient to extreme weather events, tested most recently by Hurricane Harvey in August 2017, which made landfall at the coast at San Jose Island, very close to the refuge.

Hurricane Harvey is an event that left a permanent mark on my wife and I. We both love Texas but have vowed that if we ever move back to the state, we likely won't choose to reside in Houston again because of the high risk of flooding posed by heavy rainfall events and especially powerful hurricanes, which are increasingly common along the Texas Gulf Coast.

Hurricane Harvey nearly destroyed our house in Houston. For three days and three nights, the rain simply wouldn't stop. On the worst night of the storm, we noticed floodwater rapidly rising in our street. We scrambled to alert our neighbors, so they could move their vehicles higher up their driveways as best they could. Afterward, I nearly broke my back bailing water out of our small backyard swimming pool, fearing reverse overflow from the street. My wife did her best to sandbag the front door just above the stoop. Then we both scrambled to move our most valuable possessions to the second floor, locked our cats in one of the rooms there, and steeled ourselves for several days of camping there with what supplies we had. We fully expected to be trapped on the second floor for some time as water filled our first floor, since we saw on television that thousands of our neighbors were suffering that very fate.

After bailing and sandbagging and rushing to put together our makeshift shelter, my wife and I finally managed to go to bed at something like 3:00 a.m., utterly exhausted, bodies aching. Normally, I wouldn't be able to sleep in such a situation but managed to, given my exhaustion. We were convinced that we would wake up the next morning with the entire first floor covered in a few feet of filthy water.

We were very lucky. The following morning, we woke up amazed to see the first floor completely dry, then shocked to discover a raging river flowing just a few feet from our front step, completely inundating our front yard. Our house was just high enough to stay dry, barely. Our neighbors a couple houses down weren't so lucky. We were very, very fortunate, but we will never forget that week. A year later, we sold our place and departed Texas. If we return, it will be someplace farther inland, high and dry and away from the worst flooding risks (Houston suffered disastrous floods on multiple occasions while we were living there; Harvey was just the worst episode by far).

Hurricane Harvey, a Category 4 storm when it made landfall, was the largest single rainfall event in U.S. history at the time of its occurrence. Where we lived, we received about 40 inches of rain, or roughly 100 centimeters, in just three days. Some places recorded nearly 60 inches, or almost 150 centimeters, of rain. It's Texas's costliest natural disaster, but we never lost power or even internet service during the storm, because by the time Harvey reached Houston, its winds had died down, making Harvey mainly a flooding disaster.

That storm wrought much worse damage to Rockport and other communities on the Gulf Coast unfortunate enough to find themselves directly in the storm's path and at peak wind strength. A hotel at Rockport we like—the same hotel used for our first visit to the area—was devastated and had to be rebuilt, the owners completely shutting down operations for more than a year. Much of Rockport itself sustained heavy damage, and the storm likely contributed to a steep population loss for the community.

But Aransas itself endured.

Harvey essentially passed right over the refuge at the peak of the storm's strength—the eye passed directly over portions of the wildlife refuge, causing extensive flooding and damage to facilities, roads, and other infrastructure. Aransas was closed to the public for several days following the storm, but later Fish & Wildlife happily reopened the refuge's doors, reporting that Aransas ultimately sustained relatively mild levels of damage compared to other protected areas impacted by Harvey. It was an unexpected outcome. Years earlier, Hurricane Ike all but destroyed Galveston Island State Park. Aransas was lucky, and its survival probably says something about the resilience of barrier islands and coastal wetlands.

The AWB whooping crane population was also spared—Hurricane Harvey passed over the region in August, many months before whooping cranes begin their annual arrival from Canada. Thus, the whoopers were entirely unaffected by that storm. Another endangered species native to this region wouldn't be so lucky.

Just before the environmental group TAP lost its long battle in court on behalf of the whooping crane, AWB whooping crane fans were greeted with much more positive news.

In August 2014 at a public hearing held in Houston, the Texas Parks & Wildlife Department announced what it called at the time the "largest conservation land purchase in Texas history." TPWD, other state agencies, and nonprofit organizations, in particular the Nature Conservancy, combined their efforts to secure the purchase of the 17,000-plus acres of Powderhorn Ranch, a property adjacent to the easternmost and smallest portion of the Aransas National Wildlife Refuge. It's one of the biggest developments in Texas wildlife conservation circles to have occurred since the 2010 Deepwater Horizon offshore oil drilling rig disaster. Money paid out by the companies responsible for that disaster, the largest offshore oil spill in U.S. history, later would be applied toward preserving Powderhorn Ranch and for other long-term conservation initiatives up and down the Gulf of Mexico coastline and beyond.

The Texas Parks & Wildlife Foundation initially purchased Powderhorn Ranch from Cumberland & Western Resources for $37.7 million. As mentioned earlier, the ranch is so named because it sits on the southern banks of Powderhorn Lake, directly across from the northern section of the Aransas National Wildlife Refuge and next to the community of Port O'Connor. In addition to setting aside these 17,000 acres (17,351 acres, or more than 7,000 hectares, to be precise) of land, this acquisition means that about 11 miles of additional Matagorda Bay coastline will be left undisturbed and protected from development, pollution, and other threats, ideally.

For a while, the property was placed under the care and stewardship of the Nature Conservancy before it was formally turned over to the Texas Parks & Wildlife Department. TPWD's announcement in Houston that year was widely applauded as very good news for the long-term survival of the AWB whooping crane population. In the years to come, a smaller portion of the Powderhorn Ranch will be transformed into Texas's newest public state park, made accessible with camping sites and hiking areas and whatnot, but the rest will be kept largely off-limits to the public except by permit, managed as a state wildlife management area, essentially a state-government-run wildlife refuge, operated primarily for the purpose of AWB whooping crane conservation.

I later describe my first adventure to Aransas. But jumping somewhat back and forth through time, let's look at Powderhorn Ranch in a little more detail to highlight how important this development was to whooping crane conservation—and why the powers that be made such a big deal out of it at the time.

Formerly private land, the 17,000-acre Powderhorn Ranch is a relatively undisturbed expanse of southeast Texas coastal prairie and wetlands located in Calhoun County adjacent to the community of Port O'Connor that takes its name from Powderhorn Lake. I don't know where Powderhorn Lake got its name from. Perhaps it vaguely looks like a powderhorn to some (though I don't think so). As noted, it was first placed under the protection of the Nature Conservancy and held in trust for many years by a coalition of conservation agencies led primarily by the Texas Parks & Wildlife Foundation, with the intention of purchase by the state once it had the money to do so. In line with that plan, the ranch was transferred completely to the Texas Parks & Wildlife Department's authority in late 2018.

The bulk of the land acquisition cost, $34.5 million, was made possible by funding to the National Fish & Wildlife Foundation's Gulf Environmental Benefit Fund, which is money paid out by BP and Transocean Limited per a settlement for their role in the 2010 Deepwater Horizon offshore rig explosion and subsequent Gulf of Mexico oil spill.[9] The original owner of Powderhorn Ranch intended to make it a private exotic animal hunting reserve, but thankfully those plans never materialized. That guy's mark on Powderhorn Ranch remains, however—during my investigation of the property in early 2018, I encountered several large, exotic, strange-looking Asiatic deer, specimens originally imported for the purpose of making the entire place a shooting zoo, which unfortunately actually do exist in Texas.

With the incorporation of Powderhorn Ranch, the area of legally protected habitat available to wintering AWB whooping cranes in southeast Texas is now expanding by approximately 13 percent—which isn't bad considering that the Aransas National Wildlife Refuge is already quite large. As I said, the land is located across from Powderhorn Lake next to the easternmost small subsection of the Aransas National Wildlife Refuge, thus greatly expanding the amount of legally set-aside habitat for AWB whooping cranes in the region of Indianola and Port O'Connor. The announcement was widely celebrated in conservation circles.

Today, the majority of Powderhorn Ranch is now a Texas state wildlife management area, with a smaller section slated to be developed into a publicly accessible state park at some point in the near future. The state park may be open by the time this book is published, but delays are almost certainly inevitable, since it will take time and money to develop the park to make it more accessible to campers and hikers.

I first toured Powderhorn Ranch in March 2018 with the help of a man named Richard Kostecke, who was then associate director at the Austin, Texas, office of the Nature Conservancy. My site visit revealed to me this plot's great conservation potential for the AWB whooping cranes. There's some woody brush that may need to be cleared, but overall Powderhorn Ranch is blessed with the same kinds of coastal marsh grasslands and brackish mudflats indistinguishable from those found at the nearby Aransas National Wildlife Refuge. But don't take my word for it.

A separate site survey conducted by the Bureau of Economic Geology, University of Texas at Austin, determined that the Powderhorn Ranch holds "immense" ecological value, containing within it ideal habitat suitable for a wide variety of species but particularly suitable for whoopers.[10] The study gushes that the 7,000-hectare property, though smaller than Aransas, contains within it "18 kilometers of bay frontage, tallgrass prairies, fresh and saltwater wetlands, and live oak woodlands established on a Pleistocene barrier island complex." There is ample evidence that AWB whooping cranes are poised to expand their range to this newly protected ground. In fact, I witnessed adult whooping cranes foraging near the ranch, and Kostecke insisted that he had seen them feeding at the ranch itself on several occasions.

The incorporation of Powderhorn Ranch marks the newest major addition to a broader Texas coastal conservation initiative. It also will ensure that state agencies and nonprofit groups alike will continue investing a substantial amount of time and resources to further enhancing the ecosystem there.

The majority of the former ranch will remain a state wildlife management area, essentially a state-controlled wildlife refuge where public access is heavily restricted and regulated. Researchers based at Texas university research institutions have taken an obvious interest in the site and can be expected to lend their expertise to future monitoring and assessment of the land quality, hydrology, and vegetation there. The Powderhorn Ranch and adjacent region are also covered by an existing coastal protection plan pursued by the U.S. Army Corps of Engineers, which aims to invest in land restoration, marsh restoration, breakwater development, and erosion control along an 11-kilometer stretch of the Matagorda Bay shoreline that acts as coastal storm defense for the Powderhorn Lake estuary.[11]

Most importantly, the addition of the Powderhorn Ranch marks the biggest change to whooping crane conservation since the whoopers became one of the first federally listed protected endangered species in the United States back in the 1960s, with that endangered status reaffirmed under the 1973 Endangered Species Act.

Authorities in both the United States and Canada are convinced that habitat is the key to whooping crane recovery. In other words, they believe that more and better habitat equates with more and healthier whooping cranes over time. Thus, the overwhelming emphasis on habitat management in whooping crane conservation. One study suggests that the AWB whooping crane population needs to reach a number of at least 1,000 individuals before wildlife managers can gain confidence in the species' ability to survive and thrive on its own.[12] Not coincidentally, a recovery plan in accordance with Canada's Species at Risk Act aims to see 1,000 AWB whooping cranes breeding in Wood Buffalo National Park by 2035.[13] With more than 540 individuals at last count, this suggests that management efforts are halfway to meeting that population goal.

This emphasis on habitat area is very much alive and well north of the border. Parks Canada estimates that a whooping crane breeding pair requires anywhere from 400 to 1,000 hectares of ideal habitat, mainly wetlands, to successfully breed and thrive. These conditions have almost certainly already been met in Canada—not only is it massive, Wood Buffalo National Park is a Ramsar Convention–listed Wetland of International Significance, home to a vast wooded expanse but also massive inland deltas and wetlands. Efforts to acquire additional whooping crane habitat in Texas were constrained for decades until current efforts to incorporate the Powderhorn Ranch near the Aransas refuge as additional AWB whooping crane habitat.

Below are some important points, which I get back to later in this read.

First, it's important to note that the AWB whooping crane is a fair-weather bird. By this, I mean that it spends its summers in cooler central Canada and its winters in mild southeastern Texas. Thousands of North American retirees do the same thing: flock north in the summer to avoid the worst of the summer heat, then south before winter sets in to escape the snow and bitter cold. The AWB whooping cranes have been living their lives like this ever since conservation efforts got underway, naturally migrating across a huge swath of North America twice a year, even when they numbered in only a couple dozen individuals.

That brings me to my second point: the AWB whooping crane flock displays extreme migratory behavior, completing a journey of some 2,500 kilometers from their summer breeding grounds in Canada to their wintering grounds in Texas twice a year (for a combined annual travel distance of some 5,000 kilometers, more than 3,000 miles).

Because of this migration, Fish & Wildlife officials at the Aransas National Wildlife Refuge have a window of only about three to four months to conduct the kind of population health monitoring and other activities necessary to ensure the population is healthy. For the rest of the year, the whooping cranes either spend time at various way stations in the Midwest (hopefully without getting shot in the process) or breed and enjoy the relatively cooler summers at the northern limits of Wood Buffalo National Park, where they become Canada's problem for a few months of the year.

AWB whooping cranes are also notoriously shy. Seriously shy. In fact, the best way for visitors to see AWB whooping cranes at the Aransas refuge is via boat tours departing from a dock in Rockport, but even this doesn't guarantee that one will get close enough to take a clear picture. Trust me; my wife and I tried on a few occasions to get close enough to take good pictures of whooping cranes but could manage only some blurry shots from a distance—the long-range zoom lens of her camera wasn't handy.

Another peculiar feature of this species is their unusually antisocial behavior. Though they migrate together and spend time in relatively close proximity during layovers along their migration route, presumably enjoying each other's company along the way, AWB whooping crane couples spend most of their time at the Aransas refuge clustered in their mini nuclear family units, endeavoring to separate themselves from other cranes as much as possible. Why, I'm not sure, but this is how I witnessed them foraging at Aransas, and the experts there assured me that this is perfectly normal and natural and expected for their whooping cranes. The folks running Aransas repeatedly confirmed what I witnessed: crane families mostly keep to themselves and avoid outsiders while foraging at the refuge. And they shy away from humans. I mean it; they don't particularly care for human visitors at all, not one iota.

AWB whooping cranes are decidedly antisocial. They don't like people. They don't particularly care for each other, either, unless they're searching for mates. Kind of like life in New York, when you think about it. But this behavior exists for a reason. Since humans almost wiped them off the face of Earth, they're naturally wary of us. They have good reason to be. Being shy and reticent around humans, other wildlife, and even each other ultimately may have helped them to survive extinction, but just barely.

Federal protections arguably have helped more here. And the AWB whooping crane is lucky enough to enjoy the protection of two national endangered species statutes and two sophisticated and well-funded government conservation agencies. They have a massive swath of territory at their disposal, too. This has no doubt helped them rebound from the inevitable setbacks in their recovery, including the mean Texas droughts and TCEQ's disregard for Texas's coastal ecosystem health. The whooping crane enjoys many advantages that would make it the envy of other endangered species, including Japan's red-crowned crane.

But the red-crowned has been doing even better, despite whoopers' strong position here.

Are there any pertinent points or lessons to be gleaned from all that rambling you just read through (and so very patiently, I might add)? Yes, there are a few.

For starters, some of the most important and well-known endangered species conservation initiatives long predate the Endangered Species Act itself—an important point to keep in mind. This is especially true for the whooping crane. Its most critical U.S. habitat was set aside for protection in the 1930s, and concerted, well-organized efforts toward the species' recovery began in earnest around the 1950s. Whoopers weren't officially declared an endangered species until the 1960s, a status that was later rolled over into the ESA in 1973. So a lot of what's written in the ESA was born from earlier conservation efforts, practices, and experiences. A learn-as-we-go approach to thwarting extinction with which the Fish & Wildlife Service was forced to experiment for decades played a big role in how the ESA was ultimately devised and drafted. This is also true in the case of Japan, as I explain later in this book.

Also, the Endangered Species Act empowers communities and nonstate organizations to act on behalf of listed, unlisted, and delisted species. This will continue driving much of the evolution of endangered species management in the United States for some time, if only because state agencies sometimes fight protections while federal regulators often do nothing to stop them. The folks who came up with these provisions were endowed with the keenest of foresight, I believe—they knew that on some occasions federal regulators would simply decline to enforce the law. The Aransas Project dragged Texas to court in its ultimately failed attempt to see the law enforced and the cranes protected. FWS is fully empowered to do the same thing, but the ESA's federal authorities frequently and routinely decline to do this, and the case of the whooping crane isn't the only example of this happening. Whether this harms endangered species recovery is debatable. However, there's little doubt that the FWS has drawn the connection between droughts in Texas, impacts to brackish blue crab breeding grounds at Aransas, and elevated whooping crane mortality, yet it chose to do nothing to try to rectify this situation. It probably guessed (correctly, I think) that fighting Texas on the issue ultimately would prove to be a fool's errand. FWS has limited resources, after all, so it needs to choose its battles carefully. It was left to TAP to try, and this is fully allowed under the ESA—it's even encouraged.

We also see through TCEQ's example how state environmental regulators are frequently enemies of endangered species protections, not allies. This situation could even worsen over time considering the rift in American society and the exacerbating red state/blue state divide, one that's unfortunately egged on and encouraged by our press corps. ESA implementation is heavily litigious, so the courts can be employed to compel state ESA compliance even when the federal authorities choose to not even try. It just doesn't always work out in the conservationists' favor, as TAP learned the hard way.

We also see glimpses of a phenomenon that's been confirmed by academic researchers: the longer an ESA-listed species is listed as endangered, the better its population recovery. Whooper is one of the original federally declared endangered species. It's been listed as long as U.S. federal endangered species laws have existed. And it's working—the species is rebounding, and the Endangered Species Act has clearly played a role in this. The crane's recovery has been remarkable, if very slow, especially when compared to another crane species that I know a little bit about.

CHAPTER FOUR

The Circuitous Path

What follows is a long-winded spiel about the earliest years of my career trajectory. It's probably the longest and seemingly most pointless tangent I'll go off on in this entire book, but as I mentioned earlier, there's a method to this madness and lessons buried deep in these next several paragraphs, so please bear with me.

Back when I was a beat reporter, which of course wasn't all that long ago, people occasionally asked me, "Hey, how did you get into journalism anyway?" It's a valid question, especially given the difficult state the profession finds itself in these days. Many of my childhood friends were shocked to discover I had become a reporter.

I only wish I had some sort of inspiring answer to give. I could have told how I grew up admiring those intrepid reporters who were on location when the Berlin Wall came tumbling down, or when Nelson Mandela was released from prison and apartheid in South Africa crumbled, or when George H. W. Bush's "smart bombs" began raining on Baghdad at the onset of the Persian Gulf War (I know a guy who was there). I certainly recall feeling impressed by the courage of the photojournalists risking their lives to document the exploitation of child soldiers in West Africa, and those who risked their careers to deliver justice to the victims of corporate crime. Who wouldn't? Do you remember the days of the honest, neutral, reliable, straight-down-the-middle TV news anchor, the kind you could watch for years and years and never quite figure out where they stood on the political spectrum? I do. I miss those days.

How did I get into reporting? I wish I could say I had journalist heroes or role models to look up to or that I had a relative in the business and came to know a bit about the job that way. The truth is I stumbled into journalism by accident.

I never purposely sought to become a journalist and never even considered studying journalism when I was in college. It never once crossed my mind. I think I briefly tried a mass media course during my undergraduate years but quit after one class because I found the content uninteresting. Yet one day I sort of accidently found myself in a reporting job and realized it was too late to back out of it. So I continued, staying with the industry for quite some time, right up until much of the U.S. press corps lost its collective mind and integrity, and I had to make my way toward the exit. How does one stumble into journalism by mistake? It's probably more common than we realize.

During my first graduate degree program at the University of British Columbia in Canada, students had two options to fulfill their final graduation requirements, aside from successfully completing all the necessary credit hours. One, we were given the option of writing a traditional thesis whereby we would undertake a lengthy research project, write a lengthy paper on our project or topic, perhaps (or perhaps not) get a portion of that paper published in an academic journal or graduate school bulletin, and then defend our thesis and ideas before a graduation committee. Or two, we could get a really interesting internationally oriented job or internship and then write our thesis based on that experience instead. I picked door number two and was very fortunate to choose this option.

I landed a full-time assignment at a specialized consulting firm headquartered in Tokyo. The company was founded by a guy from Brooklyn who went to Japan in the early 1990s for the sole purpose of learning Japanese. He never left and instead transitioned from freelance translation and English editorial work to going into business for himself. The company he founded fields work for several clients, mainly translating documents or even entire books from Japanese to English and vice versa, later expanding to other languages. The company also handles design, layout, editing, and other multimedia tasks. The founder was a character. For me, the most interesting thing about the Brooklyn boss was his passion for old Japanese silent films—he had a hand in preserving some notable examples and even published a book on the subject.

This company owner also took advantage of some early opportunities that popped up in his life and career to build business bridges with the Japanese government. By the time I arrived in Tokyo to join the firm, the company regularly secured contracts to handle a portion of the Japanese government's foreign media relations efforts and strategies. One of these contracts gave his company the responsibility to help organize, record, and transcribe press conferences held in English for Tokyo-based ex-pat and foreign correspondents. This work was conducted at two places: the Ministry of Foreign Affairs and the Prime Minister's Official Residence (the name was recently changed to Office of the Prime Minister, Japan's version of the White House). So this was my first job not quite out of college. I was given a pass and began working at both government agencies almost immediately, assigned to record questions and answers, identify all speakers, and then later use recordings and notes to compile full transcripts of entire press conference proceedings for speedy delivery to a government representative in charge.

The first time I managed a foreign press conference, it just so happened to be the very last press conference ever held at the original prime minister's residence building, an aging structure built in 1929 that miraculously survived the war. The Japanese government had just finished construction on a new prime minister's residence next to it, the new Kantei, and all the office functions of the prime minister were being moved to that larger, newer structure, which is more than twice the size of the classic building. The old building and its rooms are quite cramped compared to the newer one, but this didn't matter much because, as I recall it, there were only maybe five or six foreign journalists in total who turned up to ask questions to the deputy spokesperson. The master of ceremonies was a rather abrasive woman (whom I would again encounter many years later at the United Nations—naturally, she didn't remember me) put in that role not for her charm but probably due to her family connections and fluent command of English.

She started off press conferences with a couple of announcements and other brief mentions then invited questions from the floor. My job was to place microphones and recorders before the journalists and then sit quietly at a corner desk with my laptop and type up as much as I could before the press conference ended. Half the time, she couldn't answer any of their questions, but later she'd instruct me to edit the transcript to reflect an alternative reality in which she eloquently and precisely provided intelligent, insightful answers to every single question, recalling facts and details with stunning accuracy. My guess is she ordered us to fake the record because her superiors would see these transcripts later, and she wanted to impress them or at least to avoid giving them an excuse to demote her. The journalists would never get a chance to review these transcripts. They were for internal consumption only. It was actually a brilliant ruse on her part, and we went along with it.

I can't recall a single question any of those journalists asked, mainly because I was so busy. But there's another reason: I do remember that the questions were bland and largely pointless and that she mostly deflected or couldn't provide answers. Also, it seemed clear to me that the reporters present already had in mind the stories they wanted to write—they were just fishing for quotes to fill their predetermined reports. A Bloomberg journalist briefly took an interest in me, and I eventually figured out why: he suspected I might be some sort of spy or U.S. government agent. He lost interest when I made it crystal clear to him that I was just an idiot kid who had landed an interesting job.

That was my first true exposure to the world of journalism.* I came away from the experience unimpressed and began considering entering law school upon my eventual return to the States. The foreign press conferences held at the Ministry of Foreign Affairs (MOFA) were far more interesting for me.

For starters, the briefers changed, and more reporters showed up to those briefings. The questions they asked also tended to be more interesting and probing. I also liked the location better—back then, the real MOFA headquarters building was undergoing renovations, so the ministry was in temporary housing in a skyscraper in Hamamatsucho, far away from the stuffed shirts and boring vibe of the Kasumigaseki neighborhood. They let me eat lunch there, and the view was terrific.

MOFA must have been satisfied with the work that the company produced for them, because shortly after I began working with this firm, we were offered a job writing a speech for the foreign minister. She wanted to deliver an address to the Foreign Correspondents' Club of Japan announcing her official trip to Afghanistan and Iran, a state visit that had already been set up but was being kept secret from the public and press corps (Japanese and foreign alike). After being sworn to uphold this secrecy, the company president assigned me and two other employees to the task. We pulled it off after some weeks of very interesting and illuminating back-and-forth with a dozen MOFA flacks hell-bent on getting their fingerprints all over the draft—thanks to them, a well-written, quasi-inspiring speech quickly transformed into a bureaucratic afterthought of a document.†

My fondest memory from my work with the folks at MOFA was a grueling late-night assignment that kept me in their temporary headquarters until 3:00 a.m. the next day. In September 2002, the prime minister flew to

---

* Actually, this isn't completely accurate. In the late 1990s, a reporter with the *Minneapolis Star Tribune* interviewed me on the campus of the University of Minnesota. She asked me what I made of a whole host of organizational changes and new construction happening around campus all at once. I told her it was too much, like the administration was putting the student body and faculty through a type of shock therapy. My parents nearly choked on their weak American diner coffee when they read the newspaper the next day and found me quoted on the front page.

† If you're having trouble sleeping: www.mofa.go.jp/region/middle_e/fmv0204/speech.html.

Pyongyang, North Korea, in a historic first-ever visit by a top Japanese leader to the Hermit Kingdom. Why? Well, decades earlier, the regime of Kim Il Sung had kidnapped a dozen or so Japanese citizens supposedly to train spies on how to speak and act Japanese. They likely kidnapped dozens of Eastern Europeans and possibly thousands of South Korean citizens over the years, too, but only Japan's government cared to take issue with this crime against humanity. Everyone knew this had happened, but the government of North Korea vehemently denied any involvement whatsoever—until Prime Minister Junichiro Koizumi's trip.

Back then, Pyongyang was fearful of a vengeful post-9/11 United States, which had just invaded Afghanistan and was threatening to invade Iraq. There was even bullish (but mostly empty) talk in Washington of invading Iran, as well. George W. Bush's infamous "axis of evil" speech, otherwise known as the January 2002 State of the Union address, likely got Kim Jong Il thinking. Bush's State of the Union speech that year proved to me beyond a reasonable doubt that politicians have absolutely no shame and will lie about almost anything if it suits their purposes. Bush pressed the lie that Iraq was a threat to America and allied with al-Qaeda. Two of his three "axis of evil" members were, of course, Iraq and Iran, two countries that despised each other and had even fought a vicious and bloody war in the 1980s. And as you'll recall, his third member was the hapless regime in North Korea.

By this time, Kim Il Sung was of course long dead. His son feared the worst and decided that it was high time to improve diplomatic relations with the neighbors in order to ward Washington off—or at least to buy Pyongyang time until the regime could finish fabricating its first nuclear weapons. So Pyongyang reached out to Tokyo. Koizumi's government informed Mr. Kim that the only way they could make nice was by resolving the issue of the kidnapped Japanese citizens. Kim Jong Il agreed to fess up, and Koizumi agreed to hear his confession in person.

Thus, Koizumi made his first trip to Pyongyang, flying there and back in a day without staying overnight. Kim pinned the blame on daddy and promptly freed some of the kidnapping victims, at least the ones who were still alive. Questions linger about the fate of several others. Upon Koizumi's return, his administration quickly briefed the Japanese press corps, and then later a high-ranking vice minister at MOFA held a late night (close to midnight) off-the-record briefing for foreign correspondents, conducted entirely in English for the reporters' benefit. Yours truly was the guy in charge of taking attendance, setting up all the recording equipment, identifying the speakers, and then later transcribing the entire thing for the Japanese government's records. It may have been closer to 4:00 a.m. by the time I got home, completely exhausted, courtesy of a MOFA hired car. Thankfully, Brooklyn boss gave me the following day off. It was quite an experience, and I easily graduated and earned my degree after putting this all down on paper for my adviser and graduation committee.

Brooklyn boss later asked me to join his firm permanently upon graduation, making a very generous offer to entice me to stay, but I very reluctantly (and foolishly) turned him down and instead flew back stateside about a month later, settling in Seattle. By then, Bush's economy was in a serious rut. But after some perseverance, I eventually found employment in the only industry that was actually experiencing growth. America's looming economic catastrophe, the Great Housing Bubble, was just finding its legs, and I found accounting work with a house builder in a Seattle suburb.

However, and most fortuitously, before I left Tokyo, I dropped off my résumé with a headhunter, actually a hiring agency specializing in landing work for bilingual professionals at Japanese companies. I noted in my file that I was open to assignments in both Japan and the United States, guessing that I would get the greatest number of leads that way. It was a good thing I registered with the agency, because some two years after my departure from Tokyo I received a phone call out of the blue asking if I was still looking for work. At that time, I was most interested in a change of scenery.

"I see you have some experience with the media," the headhunter told me (my résumé listed the press conference work). "You majored in international relations, with a focus on East Asia, yes?" Correct, I did. "Also, it shows here translation experience, is that right?" Indeed, I did that, too. "And you must have some familiarity with research since you went to graduate school, yes?" Yes again; indeed, research goes hand-in-hand with grad studies. "I have an interesting job opening I thought you might like to consider," she said. "I'll email you all the details." And she did. "Please look over the job description carefully and let me know if you are interested," she asked. I was.

Nihon Keizai Shimbun, more familiar to Westerners as Nikkei, is Japan's largest financial news service and quite likely Japan's wealthiest media company. It's expanding globally these days, having recently acquired the *Financial Times*. Nikkei's Western Hemisphere headquarters is in New York. It shares an office tower with Warner Brothers (I once ran into the actor William Fichtner of *A Perfect Storm*, *Black Hawk Down*, and *Prison Break* fame in the lobby) and other notable companies on the Avenue of the Americas, very close to Rockefeller Center. The company was hiring and was apparently either having trouble filling the position or keeping employees in that assignment, which is why this headhunter reached out to me with some urgency. She seemed desperate to get Nikkei qualified candidates for interviews, even if it meant flying them in all the way from Seattle.

The job description she emailed to me said, "Researcher, United Nations" and simply explained that Nikkei wanted someone to perform regular research functions for it at UN Headquarters in Turtle Bay, on the east side of Manhattan several blocks away from its Avenue of the Americas office. From the job description, it sounded like I would be spending most of my time in the UN's library or at various Secretariat offices compiling statistics on global economic conditions, which would then be folded into Nikkei reports. That's all it alluded to. "Researcher, United Nations" was all the ad said. By this time, I was already beginning to mull the idea of going back to school to try for my PhD, and I desired some real-world, practical work experience that would translate well to that, ideally of the international variety. So here was my chance—move to New York to conduct research on the world economy and global society for a financial news company. I told the headhunter that she could go ahead and send them my résumé. About a week later, they flew me to New York for an interview. I took the red-eye from SeaTac Airport.

I felt this was my last opportunity to land this type of work. A couple years before all this happened, I flew to Washington, D.C., one summer day to interview for a job at the State Department. It was my very first horrible post-graduation job-hunting experience. They made all of us candidates do this stupid group exercise where I proved to them how bad I am at bureaucracy and how I was all too reluctant to waste taxpayers' money on meaningless things (a cardinal sin in D.C., apparently). Not a single one of the candidates in my group passed, so afterward we all went out to a bar together to get drunk (the pub was called Hawks and Doves or something to that effect). One of my failed co-applicants was a lawyer looking to switch careers. She entertained us with a fascinating tale of how her ex-husband had been indicted by the FBI a week after their wedding. He'd been laundering money for some rather nefarious individuals, as it turned out, and the feds spent the better part of a year building a case against him. Naturally, this came as a shock to her, so she used her lawyer powers to get their marriage annulled. She had a wild streak—that entire night she played footsie with me under the table and tried to lure me back to her place. Washington is an interesting town.

Anyway, that trip to D.C. ended up as a horrible job-hunting experience capped off with a horrible hangover. Things didn't go much better for me in New York, either. In fact, it was a disaster.

I was greeted by a huge billboard in Times Square with a smiling picture of comedian Jon Stewart. The caption next to him read—and I'm not kidding—"Welcome to New York! That smell? Freedom." It was the perfect welcome sign—New York really did stink that day. It was hot, sticky, oddly smelly, and just generally horrible. Remember, I had just spent half a year working in Tokyo. Tokyo isn't particularly easy on the eyes, but it's generally clean and well-kept, ordered, and tidy. Then there's New York. You probably know what I'm talking about. Nikkei's office was nice enough, but things only slid further downhill there.

The bureau chief and a senior reporter, the would-be researcher's supervisor, decided to opt for a group interview—the worst kind of job interview. I was paired with an arrogant prick freshly graduated from Harvard and some other guy, a local. The interview was held in Japanese because the bureau chief couldn't speak English. Normally, this wouldn't be a problem for me, except the other two applicants spoke terrible Japanese. I couldn't make out a single word they said, and they were always prompted to speak before me. So, of course, they dragged me down to their level, making my Japanese terrible in the process as I followed them in answering a series of general questions. The interviewers' faces said it all. I later stepped out to the street, extremely confident and relieved that I wouldn't be getting any job offer from Nikkei that week, so I made a beeline for JFK Airport. I wasn't supposed to fly out until the next day, but I just had to get the heck out of New York City as quickly as possible. I lucked out—there were available seats on the next flight to Seattle, and a gracious and polite attendant agreed to change my ticket for no charge.

Three months later, the same headhunter called me again. "Are you still interested in that researcher assignment at Nikkei?" "Well, yes, I suppose so," I said (I really wasn't, but the job in Seattle was getting gradually less tolerable). "But they went with someone else, didn't they?" I asked. "Yeah, that last guy didn't work out," she explained—he quit abruptly. She asked me if I would be willing to fly out again for a second interview. "Things might go better this time," she pleaded with me. "Sure, I might as well," I answered. I wasn't relishing the idea of moving to New York, but I needed something better for my career. I was feeling stuck in the Pacific Northwest. So I booked another red-eye for the Big Apple. The second interview went much better than the first.

The second time around, it was just the three of us—me, the bureau chief, and the senior reporter. No other applicants to throw me off. I nailed the Japanese, impressed them with my answers and credentials, and assured them that researching statistics and international social and political conditions from the United Nations was well within my capabilities and skill set. I also wouldn't be intimidated in that environment, given my past experiences in Tokyo at Japanese government offices. They offered me the job on the spot. I had to start in a week. Shortly afterward, I began growing suspicious of what I had just signed up for. Researcher, United Nations, right? That's what the advertisement said. I read the job description over and over again and never came to any other conclusion. And that's how they explained it in both interviews. It sounded like good ole academic library work, lots of reading, occasional queries to UN staff, report drafting, tons of writing, etcetera, etcetera, tasks any other person applying would have reasonably expected. Perfect training and preparation for PhD studies, I thought.

My very first assignment with Nikkei was to find and interview surviving family members of the victims of the 9/11 atrocities in time for the third anniversary of the terror attacks. Ideally, children and widows, per my new boss's instructions. This, of course, had nothing to do with the United Nations. Naturally, I was a tad bit confused. Not to mention, I was assigned this task prior to my move from the West Coast. But I did what I was told.

It was only after relocating to New York from Seattle that my supervisor helpfully informed me of what my actual job title would be: assistant reporter, Americas Bureau. The "research" he had in mind was for me to jot down notes and quotes and information concerning various breaking stories occurring at the UN and up and down the Eastern Seaboard. These details then would be worked into news articles that he would write in Japanese. This senior reporter hated the UN and didn't want to spend any time there personally, so he crafted the advertisement in a way to entice someone who could largely do all that's necessary to generate a news story independently with only light supervision, while he called most of the shots from his office in Midtown. He didn't want to recruit a journalist or a journalism major for the job, either. I think because he feared that might cause trouble or friction for some reason. But as I quickly discovered, there would be no academic-style "research" to be had. I was now a journalist, charged with helping another journalist more senior than I write my stories in Japanese. Like I said—stumbled into it by accident.

Eventually, I found my desk at the United Nations, where I would spend the bulk of my time following the activities of the UN Security Council. I think I stepped foot in the UN's library maybe once during the entire seven years I spent there employed as a resident correspondent. There were other interesting assignments, too. They flew me to Boston to watch John Kerry lose to Baby Bush in the 2004 presidential election. I arranged two photo shoots with New York Yankees outfielder Hideki Matsui. When poor Mr. Matsui and the Yankees lost a division series to the Red Sox and thus a chance at the World Series, my boss, who happened to be friends with Matsui, had me book two airline tickets for Matsui and his then actress/model girlfriend to the Caribbean to help him take his mind off the defeat, apparently. They figured having me book the flight and hotel would circumvent Japan's paparazzi, and it worked—the tabloid media didn't find out about any of it until after Matsui and his gal pal returned. Later, I interviewed Woody Allen at a fancy Manhattan hotel and then later still baseball legend Hank Aaron in Atlanta (thanks to my wife, whom I met at the United Nations). And, of course, I spent many hours and days following world leaders' speeches, pitching questions at diplomats, sneaking around at night looking for secrets and scoops, and far, far too much time chasing North Koreans through the streets of New York after Kim Jr. finally made his bomb and detonated it for all the world to see.

So that's how I accidentally became a journalist. I was told on my first day in no uncertain terms that I was a reporter, not a researcher, and being a reporter meant finding news every day, because the paper can't run blank. Since I had just upended my entire life and moved everything I owned to the East Coast, I figured it was too late to back out of it, even after I sorted out why the previous employee lasted only three months—the new boss was a hard-ass who proved very difficult to please. But I survived. Realizing the opportunity before me, I used my position and connections at the UN to begin freelancing work on the side, first for a now-defunct Canadian magazine, then with the Canadian Press newswire service, and later with *The Economist*. A search for interesting and mostly neglected UN news led me to focus on environmental matters, including international environmental diplomacy and comparative environmental policy. I hadn't intended to get into this line of work but managed to trip into it anyway. Sometimes, life works out that way. And here I am, boring you with all these details.

Fun times.

Employment as an international environmental journalist and the circuitous path I took to land that position were excellent training for the kind of work I mostly do today. Although I still educate college students and conduct academic research, I also dabble in environmental and natural resources news coverage, but only of the sort that I'm passionate about. Once you start writing about environmental concerns, you can't ever really stop, no matter where you might find yourself in the world. Good journalists know how to stay curious and to always be on the lookout for promising stories that are too often ignored by the broader media herd. Good environmental journalists take this attitude and curiosity to a whole other level, discovering major events or controversies unfolding amid normally mundane and overlooked daily realities, like abandoned lots, or a stream with some oddly colored substance floating in it, or a little-noticed piece of proposed legislation ignored by political reporters but of apparent great interest to lawyers and contractors. Or a floundering endangered species.

Here's another example—brownfields.

New York was experiencing something of an economic renaissance during my days there, but I occasionally ran across abandoned lots in the middle of prime real estate, bizarrely ignored by developers. After some digging, I discovered that New York had a major brownfield problem.

For readers unaware, in the United States a brownfield is a partially polluted parcel of land that by law must undergo environmental remediation before any new development can take place there. If a plot of land is polluted enough, then it can be declared a Superfund site and attract federal remediation dollars and assistance from the Environmental Protection Agency; if not, then the site is declared a brownfield and it becomes a state government problem instead. New York City was home to thousands of state-managed brownfields when I began looking into this problem.

Back then, the State of New York didn't take much interest in fixing New York City's brownfields, then a vast patchwork of abandoned lots and polluted properties. To get around this, on August 5, 2010, the city council established America's very first municipal brownfield cleanup program in an effort to speed up environmental remediation work and to boost the city's economy with new development. It was the first time a city government ever stripped a state government of its brownfield environmental oversight responsibilities, made possible only because Albany didn't care and was happy to part with the responsibility. This landmark event received almost no press coverage—even the local papers couldn't care less at the time.

Still, several months after this first-ever municipal brownfield program was launched, the city had very little to show for its efforts; these eyesores lingered on, untouched. The problem seemed to be a lack of funding, weak financial incentives for developers, and a mile of red tape that had to be cleared before any proper environmental cleanup even could begin. It also didn't help that these polluted properties were concentrated in poorer neighborhoods dominated by minority populations. I wrote about all this and in the process earned some street cred with the types of folks who pay attention to these kinds of bread-and-butter environmental issues.[1] That reporting work was aided by earlier experience I gained investigating a largely unnoticed federal government effort to strip New Jersey of some of its Superfund site reclamation oversight functions, because New Jersey's government had been dragging its heels or ignoring ongoing pollution problems for decades.[2] I stumbled across that story after growing curious about large swaths of undeveloped properties along my train route to and from the city.

The lesson here for journalists—and ecologists, environmental scientists, ecosystems researchers, or what have you—is that there are plenty of interesting stories to dig into if you know where to look and care enough to look for them. Don't simply follow the herd. Don't feel compelled to direct your attention someplace just because everyone else is, too, or divert your eyes from some question or topic because no one else is paying attention to it. When everyone else zigs, you zag—that's one key to becoming a successful journalist or researcher, the kind that a variety of folks from a wide swath of the public with a wide array of varying political opinions will come to respect.

As fascinating as you-know-who is, most reporters are obsessively covering him only because their peers in the profession are doing the same. This is the herd mentality at work, the same phenomenon that's unfortunately dividing far too much of American society, polity, and media in this modern era, splitting the nation into warring camps. It needs to end.

Perhaps New York City could have resolved its polluted brownfields problem by now if more folks had cared to pay attention to them. Perhaps conservationists might find clues about how to reverse some species' march toward extinction by having the courage to turn away from more mainstream lines of inquiry, training their lenses instead on mostly ignored or overlooked examples of where this has been achieved. It's at least worth a shot. Or a curious researcher might stumble onto some fascinating story or interesting discovery while taking a circuitous path toward another intended goal, whether that original goal was ever attained or not.

If we want to start making a serious dent in the global biodiversity loss and extinction crisis, a great first step is to begin at least paying serious attention to it. The 50th birthday of the Endangered Species Act seems as good an opportunity as any to do this. But I'm willing to bet that that's not really happening as you read these words right now. More likely than not, the clickbait and rage porn are still winning the day, and public attention spans are still focused instead on some other real and many more imagined problems that the press corps deems more worthy of covering, because these are the stories that the rest of the media herd is rushing off to follow.

But there are better, more interesting, and far more consequential stories to tell. Scratch the surface a bit more deeply, and you'll find them. You may even stumble on them by accident, but only if you let yourself.

Academics could perhaps use this tip on occasion too, I believe, but maybe in the opposite direction.

Their research often yields impressive insights and details, but too often academic research overlooks or doesn't recognize the value in scrutinizing some other interesting aspects or facts in a broader sense, maybe by taking the view from 30,000 feet, so to speak, or by simply taking their eyes off their work for a moment to see what else is going on around them. Great insights probably could be achieved if researchers had more courage to direct their research skills away from the popular minutiae to dive into some fairly well known if seemingly boring realities, such as comparing two famous conservation case studies side by side. Curiosity causes many researchers to zoom in on a topic or problem, but they sometimes forget to zoom out, or to even just look around to see if there are other similar issues or problems connected to their investigations that warrant further attention, if only for comparison's sake.

The whooping crane of North America has received an enormous amount of scrutiny from wildlife researchers and the conservation community for decades. But near as I can tell, for more than 70 years since the most active efforts to save this species have been underway, no one thought to look at a nearly identical story happening simultaneously in Japan to see if there were any lessons or insights that could be gleaned from that example, tidbits that might prove valuable to the American crane conservation effort. For 70 years—again, only as far as I can tell—no one thought to compare whooping crane and red-crowned crane conservation efforts together, side by side, to see where that exercise might take them.

I must confess: I hadn't thought to do this either after I moved to Texas and enrolled in a second graduate degree program. But I did grow plenty curious about the whooping crane, this iconic endangered species that I knew far too little about back then. By this time, my main duties involved researching and writing about the oil and gas industry—I started my energy industry coverage by following commodities markets and renewable energy finance in New York—but as I noted above, it's difficult to stop writing about the environment and conservation once you start. I kept finding excuses that would let me keep one foot in that world. Thus, I decided to investigate the status of the whooping crane and the legal battle surrounding this species, which was another overlooked but fascinating story that had been receiving far too little attention in the national press at the time (actually, practically no attention at all back then).

I was determined to experience my first encounter with the whooping crane, North America's largest bird and largest species of waterfowl, as a tourist might, rather than from the point of view of a researcher, conservationist, Fish & Wildlife Service official, or even a journalist. After all, this is how the vast majority of folks first witness these impressive creatures in the little corner of Texas they call their winter home—most individuals who spy whooping cranes are tourists, including the thousands of individual birding and nature enthusiasts who traveled far and wide to see the amazing and beautiful whooping cranes before the pandemic struck.

So one fine and sunny winter day (springtime, really, in Hokkaido terms when one considers the comparatively mild weather), my wife and I loaded our little Ford Escort with her cameras, her photography equipment, and some other basic supplies, and we set out early for the three-hour drive from our house in northwest Houston, Texas, to the Aransas National Wildlife Refuge, which can be found not too far from the interesting city of Corpus Christi. Technically, the refuge is located in the small southeastern Texas community of Austwell. An alternative launching point is the lovely coastal vacation spot of Rockwell, a community that came close to being wiped off the map only a few years later when it took a direct hit from Hurricane Harvey.

The drive itself is fairly uneventful. This part of Texas is flat and dominated by farms, which makes the trip seem longer, but otherwise it's lovely and relaxing if you play your favorite music or perhaps an audiobook during the drive. Maybe even a podcast.

The traffic on Highway 59 usually isn't too bad. You exit the beltway at Sugar Land and eventually start slipping out of the hectic mess that is traffic in metro Houston at about Rosenberg. For me, you've officially departed the orbit of Houston by the time you pass the town of Beasley, which I consider a landmark only because of a sign pointing to an exit for Grunwald Road. Perhaps some long-lost distant relation of mine farmed there back in the day. I'll never know as I've never actually taken this exit, which is fine by me, since Google's street-view photography shows mainly more farms and some farm equipment dealerships, which sums up what you'll encounter nearly everywhere along the route to Rockwell.

The scenery transforms into more scrubland and bushland by the time you reach Victoria, and it's here where you also begin encountering signs of the oil and gas industry, including—back then—what was left of the hydraulic fracturing boom that for a time dominated a vast stretch of southern Texas, overlaying something called the Eagle Ford Shale formation, a belt of geological shale oil and natural gas deposits spanning just east of San Antonio, all the way to the border with Mexico. Highway 59 turns into Highway 77, which leads you farther southwest until you hit Refugio, which is a good place to veer left and head toward the Gulf of Mexico. You can also take a left at 239 and make your way to the small community of Tivoli and then Austwell. A small county road that hugs San Antonio Bay then takes you to the main entrance of the Aransas National Wildlife Refuge, one of the largest wildlife refuges in the U.S. Fish & Wildlife Service's portfolio, at least in the Lower 48.

Unless there's some major event underway, the atmosphere here should be nice and quiet. It was very peaceful and relaxing when we first paid a visit. It seemed like we were the only people there until we reached the visitor center and saw two other cars parked out front alongside the FWS vehicles.

The surroundings were dominated by coastal prairie and patches of bushes and scrubland. It reminded me somewhat of the high desert of northern Arizona, except Aransas is much wetter, so more vegetation. The visitor center is nice enough, not much different from what you might find at other federal parks. When you walk in, you're greeted by an array of dead stuffed animals and nature displays, sort of like a mini natural history museum. These are to help the refuge fulfill its educational mission—school groups routinely visit, so a lot of the displays are designed to appeal to children and to give them a chance to experience nature up close. I found the displays beneficial, too, since I was about to discover just how difficult it is to actually encounter wildlife at this refuge. The lobby of the visitor center also houses a dead and stuffed specimen of a whooping crane, a rather young one by the looks of it (at least the one I saw).

We later drove around the refuge, hiking a bit, stopping at some nice viewing platforms to take a few panoramic photographs. It was nice—it's not Yellowstone, of course—but as I mentioned earlier, Texas has a dearth of open public spaces where people can just get away from it all. Aransas is one of them, and if you're in the area you should definitely pay a visit.

We spent the entire day there. We even saw whooping cranes, though they were so far away that it was difficult to distinguish them from egrets or any other waterfowl. I had to use a pair of binoculars to confirm that they were indeed whoopers. They must have been a good kilometer away from us, but the short vegetation and their height made them visible to us. We were visible to them, too. We tried walking closer. They reacted by moving farther away from us, despite the already huge distance between us and them. I asked the FWS official at the visitor center about this behavior. She laughed. "Yeah, they're shy birds and don't much care for humans," she said. "We can hardly get within a mile of them before they start taking off in the other direction." FWS must have discovered ways to get around this—several cranes are banded and fitted with radio collars to allow wildlife authorities to track the flock along its migration.

Maybe conditions just weren't great when we paid our first visit. Other travelers might have better luck. FWS has set up a series of tall observation towers along the coastal parts of the refuge. They're fitted with coin-operated viewfinders that let you peer at distant objects, the sort you find at any type of outdoor tourist attraction, like the top of the Empire State Building or the Grand Canyon. You might be able to spot a feeding whooping crane that way. But even if you don't, a visit to Aransas is recommended any time of year. It really is lovely.

"You should try the boat tour that departs from Rockport," that FWS officer helpfully advised. "You'll usually have better luck seeing whooping cranes up close that way." We took her advice during our second visit to the refuge, which didn't happen until the following year due to my work duties and the fact that whooping cranes spend only a few short winter months at Aransas before making their way back north.

The followup boat tour from Rockport didn't go exactly as I had hoped, either. The weather was excellent. The company, lovely (mostly retirees with the money and free time to travel to see birds in Aransas; I'm pretty sure we were the youngest pair on the boat). The captain and his crew were gracious, generous, and more than willing to point out the sights and guide us as best they could. The whoopers, however, were in much less generous a mood. Though they normally spend their days hugging the coastline searching for blue crabs, for whatever reason, the Aransas whooping cranes decided to forage farther inland the only day we were able to get a ticket on that tour boat. It's like they knew we were coming.

We got some OK shots but still were too far away to avoid blurriness. My wife brought along an impressive zoom lens, impressive to me anyway, but it wasn't enough. We could tell they were whooping cranes, but that was about it. One good thing we had going for us is that they didn't shy away or escape at the mere sight of our boat. Apparently, they're OK with vehicles, but not so much if you try to approach them on foot, as the captain explained to us. That said, the whooping cranes didn't bother to get any closer to us, either. They just sort of stared at us from the distance, and we at them. It was a deflating experience, but I got to see a bit more of the refuge from different angles.

The highlight of that trip was actually watching a heron swallow a fish that was larger than its own head. Our boat motored past a bend, and to our left we spotted it: a beautiful adult heron wading in the mudflats, with the distinct blue plumage sprouting from the eyes. We had seen herons plenty of times around the Houston area—lots of snowy egrets, too—but never quite this close. It appeared startled at first but then mostly ignored us, apparently already used to the sight of a bunch of strange apes floating past on a giant contraption.

The heron had a massive fish in its mouth. No way it can get that down, I thought. My wife thought differently. So we made a friendly bet. The captain slowed the motor as if he had sensed the wager underway. The blue heron struggled for a time with the massive, heavy fish resting sideways in its beak. But after a few minutes of just sort of chomping at it, the heron finally figured out its strategy. With a flick of its neck, the heron flipped its catch into the air just a few inches and then deftly caught it headfirst, the fish's tail sticking straight up in the air. The heron then worked the entire thing through its gullet and into its belly in less than five seconds. My wife won the bet.

Later, on another occasion, I traveled alone to the vicinity of Aransas to tour and learn a bit more about Powderhorn Ranch, the conservation community's newest gift to the whooping crane. This time I was very much wearing my reporter's cap—no tourist would have been permitted on the property, and I was back then expanding my investigation and reporting of the Texas Aransas Project lawsuit and the controversy surrounding whooping cranes and surface water management in Texas (controversial to me, at least).

The Powderhorn Ranch can be found on the northeastern fringe of the Aransas National Wildlife Refuge next to the refuge's smallest subsection between the RV park/ghost town of Indianola and Port O'Connor, a small touristy town known for its recreational fishing opportunities. I accessed the ranch from northwest of Port O'Connor off a small farm-to-market road, following Richard Kostecke's directions to the main gate, where I rendezvoused with him just before 11:00 a.m. We left my car parked at the gate and took his truck, since back then the access roads were either in poor shape or nonexistent.

I keep calling it Powderhorn Ranch because that's what it was when I visited the site, still under the care of the Nature Conservancy, with Kostecke acting as the local agent in charge of managing the massive property. He had the keys to the gate, and I required his permission and escort to get in and see for myself whether whooping cranes might feel at home on this plot of land. The Nature Conservancy had no intention of holding on to the property—it was just keeping it in trust until it could be handed over to the Texas Parks & Wildlife Department (TPWD). Eventually that's what happened—today Powderhorn Ranch is known as Powderhorn Wildlife Management Area, essentially a state-run wildlife refuge that is mostly off-limits to the public. The TPWD website says the area is "open only on specific days for hunts and bird tours." When I paid my visit, access was even more restricted. Aside from the previous landowner and his family, perhaps only a dozen officials had ever set eyes on the site, mainly to scout it for potential acquisition. I'm pretty sure I was the first journalist to step foot on the property (though I could be mistaken—Kostecke didn't say).

The tour was brief—only a couple hours—but we covered the entire property from the main access road all the way to the coast, where one can still spot the wreck of an abandoned boat rusting on the beach. I would have guessed Hurricane Harvey washed that rusting hulk back out to sea, but nope, a Google Map satellite photo dated 2022 shows that the boat wreck is still there, marooned on a sandbar, rusting away.

Powderhorn Ranch would make any wetland or shorebird happy. I spotted a massive jackrabbit. Later on, we ran into some very odd-looking deer, much larger than what you might expect to find elsewhere in Texas—or anywhere in North America, for that matter. They were nearly as big as elk but not quite and definitely not elk (which I've encountered numerous times in northern Arizona). "Those are Asiatic deer," Kostecke helpfully explained. He told me the story.

The previous landowner was an eccentric millionaire who grew jealous of the shooting zoos that Texas legally allows and that other rich Texans operated. Again, as I explained earlier, these are basically open safari camps where unethical hunters pay to shoot animals that have no means of escape, in the most unsportsmanlike way imaginable. Don't think Kruger National Park; Texas's shooting zoos are far smaller and far more densely populated than any natural game reserves—they are artificially constructed, packed concentrations of captive wildlife designed to entertain people who enjoy killing helpless creatures. Powderhorn Ranch's original owner wanted one of his own, so he imported exotic Asiatic deer with the idea of letting them breed, multiply, and then die as he enjoyed the profits. Either they didn't breed as quickly as he had hoped, or something else went awry, because the Powderhorn Shooting Zoo never got off the ground. Conservationists eventually acquired the property instead, inheriting the huge, odd-looking deer in the process. They were spending their days milling about the ranch, minding their own business, mainly keeping to themselves when I was there.

This seemed to me, at the time, like the next budding invasive species disaster in the making and on property ostensibly set aside to ensure the survival of a native endangered species. No one thought it odd or problematic, however. As far as I'm aware, the exotic Asiatic deer of Powderhorn are still there and aren't deemed much of a threat to the ecosystem or the whooping cranes.

Minus the strange deer, Powderhorn Ranch is indistinguishable from the Aransas National Wildlife Refuge: The same coastal grasses and bushes. The same eddies and wetlands. The same coastal brackish mudflats almost certainly hosting the same Texas blue crabs that whoopers absolutely love. Some very nice beaches here and there, and the same confluence and interactions between fresh water and seawater. Though I did spy some whooping cranes foraging deep in farmers' fields on my drive in, I didn't spot a single whooping crane during my entire investigation of Powderhorn Ranch. Kostecke assured me that a few whooping cranes do spend quite a bit of time at the ranch—he had seen them there with his own eyes, he swore. And I believed him. There is every reason to believe that the whooping crane is poised to colonize large swaths of Powderhorn Ranch as neighboring Aransas gets a bit too crowded for the birds' comfort. But don't take my word for it—as I mentioned earlier, a more competent group of professionals undertook a survey of Powderhorn Ranch a few years after my visit and reached the very same conclusion.

One day, you will be able to freely visit and explore sections of Powderhorn without any special hunting permit or other prior appointment. But that day won't be soon. Unfortunately, we could be forced to wait another eight to ten years before Powderhorn State Park is opened to the public.[3]

Obviously, I had read about the Aransas refuge and its most famous seasonally resident species, the whooping crane, well before I set out to tour either the wildlife refuge or Powderhorn Ranch. I was a journalist, after all, so I conducted plenty of research ahead of time. It was the same story when I first dug in to the tale of the red-crowned crane of northern Japan.

Still, I first approached these two cranes as a tourist might. It was only later that I conducted my investigations in accordance with my duties as a journalist and academic researcher, undertaking deeper preliminary investigations ahead of time while getting ready for site investigations and interviews. I did my best to dispassionately investigate these two crane species case studies, approaching both the whooping crane and red-crowned crane stories mainly through literature reviews and trained observations using some of the skills I'd acquired in graduate school and in my years as a beat reporter. Throughout this book, you might notice how I tend to jump back and forth between these different hats or points of view—touristy one minute, journalistic the next, wonky and analytical the following moment—but that's a consequence of how I've been traveling with these two species for years now, ever since my very first journey with my wife to the Aransas National Wildlife Refuge.

During subsequent years, I greatly expanded upon that initial inquiry and later did the bulk of my research into these two cranes' stories and conservation histories in Japan after I started my PhD program. But it all began with a nature lover's simple curiosity, then as part of a reporter's investigation into an environmental lawsuit, and finally as the subject of an academic dissertation and a book author's topic of choice. I toggle between all three modes in this book, so apologies ahead of time for any confusing juxtaposition you may encounter.

My initial three attempts to get up close and personal with a living whooping crane—or ideally several whooping cranes—all fell flat. First, they would have nothing to do with me as I tried to approach by land. Next, they would have nothing to do with me as I made my approach by sea. Finally, they were absent entirely when I visited what is to become their newest seasonal home in Texas. Disappointing? Sure. But no matter—I've seen them (though at a distance), they exist, and I know damn well what whooping crane conservation looks like and what it entails in North America, especially in Texas and at their migratory way stations in the Midwest.

We can't blame the whooping crane for being a shy bird. As much as I and this book's readers may cherish and value them, we may be outnumbered by the hordes who happily would wipe them off the face of Earth given even a fraction of the chance. A few years back, a guy in Louisiana set out one day to deliberately shoot whooping cranes after he learned that they were feeding in the area, knowing full well that whooping cranes are a critically endangered, federally protected species, and that killing one, even accidentally, could land you in prison for a while or greatly lighten your bank account, or both. Knowing all this, he deliberately tracked down and shot two whooping cranes, just for kicks, then left their carcasses to rot in the sun. He was eventually caught and convicted and forced to pay an $85,000 penalty, a record fine at the time, I believe. They also put him on probation for five years, banning him from hunting anything until he finished a minimum of 360 hours of community service. The judge said that the only reason the poacher wasn't given prison time was because the novel coronavirus was then spreading quickly through Louisiana's prison system.[4]

Sounds harsh? Apparently, this well-publicized record fine for killing an endangered species wasn't harsh enough. A little more than a year after that fine was handed down, a man in Oklahoma shot dead four whooping cranes.[5]

These are hardly isolated cases. At least 21 endangered whooping cranes were shot by hunters between 2011 to 2016. Most of these killings happened to whoopers that were not a part of the Aransas-Wood Buffalo migratory flock. A 2019 academic study says that 86 percent of the birds killed were targeted by hunters operating further east, where conservationists have spent decades and millions of dollars trying to revive the cranes through captive breeding and reintroduction programs. A lot of times the killing is just careless stupidity—the hunters don't know what they're shooting at but shoot anyway. In other cases, like the one described earlier, whooping cranes are killed purely out of malicious intent. One study I read described a case in which a teenager was convicted of killing whooping cranes from an extremely rare and almost certainly doomed nonmigratory population in Florida in 2000. He shot two of the cranes from his pickup truck and then casually drove away. He was too lazy to even get out of his vehicle.[6]

So the whooping cranes are well within their rights to steer clear of the likes of me or any other bipedal ape who happens to get too close. The Fish & Wildlife Service says habitat loss played a big role in the near extinction of the whooping crane, but wanton, unrestricted, careless, and callous hunting by the sorts of individuals who enjoy killing big, slow things has taken a massive toll as well.

Beginning in the early 1900s, conservationists grew convinced that the whooping crane's extinction was a foregone conclusion. This remarkable species, the largest bird in North America, was once common throughout much of the West and Midwest and easily could be spotted at wetlands and shorelines up and down the Eastern Seaboard. By 1913, experts had declared whooping cranes gone for good in Arkansas, Delaware, Georgia, Idaho, Indiana, Iowa, Massachusetts, Michigan, Minnesota, Missouri, Montana, Nebraska, New Hampshire, New Jersey, New York, North Dakota, South Dakota, Texas, Wisconsin, and Wyoming.[7] Obviously, early conservationists were incorrect about Texas; Aransas and the northern fringes of Alberta became the last refuges for slightly more than a dozen surviving whooping cranes. Early 1900s Texas hunters would almost certainly have annihilated those few remaining individuals, as well, were it not for federal protections.

Today, the wild migratory flock, the AWB whooping cranes, numbers slightly more than 540 individual birds. To put this in perspective, that's roughly equivalent to the number of California condors believed to be currently surviving in the wild. There are probably around 800 existing whooping cranes in total, including the AWB flock, much smaller flocks in Louisiana and Wisconsin, and all whooping cranes in captivity. Knowing this, the Fish & Wildlife Service apparently floated the idea of removing the whooping crane from the list of endangered species in late 2021.[8] And this under a Democratic Party–controlled administration. Obviously, conservationists are vehemently opposed to any changes to the whooping crane's endangered species status. Personally, I don't blame them—this species is still remarkably rare and difficult to spot, even when you know precisely where to go and how best to approach them.

I promise you, should you travel to northern Japan for the purpose of spotting a Japanese red-crowned crane, your experience will be far different from what I went through in Texas.

Though they aren't necessarily gregarious, red-crowned cranes aren't exactly shy either. They may look nearly identical to whooping cranes, but their behavior toward humans is quite different.

Nope, *tancho*—the red-crowned crane—isn't a shy bird by any means, at least compared to the reclusive, antisocial whooper.

My first encounter with the red-crowned crane of eastern Hokkaido, Japan went . . . well, you could describe it as a far more intimate and up-close affair than my earlier futile attempts at getting a close look at an Aransas whooping crane. I almost hit *tancho* with my rental car.

CHAPTER FIVE

The Grateful Crane

The lovable, adorable, cuddly koala—Australia's mascot—is now an endangered species. I bet you didn't know that. Or maybe you did.

That designation was made official and announced February 2022. This news caught my attention because just a month prior, a student in my ecosystem science fundamentals class delivered an outstanding presentation on the destruction wrought by two successive years of mass wildfire outbreaks in the Australian bush. Her presentation featured the koala because, obviously, that species is an iconic symbol of Australia (it exists only there natively, evolved to subsist on the eucalyptus trees, which are toxic to other species) and because she learned during the course of her research the sad news that thousands of koalas are believed to have perished in those wildfire outbreaks. During the question-and-answer portion of her presentation, I asked her if the fires resulted in the koala becoming an endangered species. She answered that although the koala officially hadn't become listed as endangered yet, she expected the authorities down under to make that move soon. She proved quite prescient.

Despite its cherished status as a national symbol down under, koala numbers had been in decline for some time before the listing decision was made, "a combination of land clearing and fragmentation, disease, climate change effects, and deaths by vehicles and dogs," as University of Sydney associate professor Matthew Crowther explained in a press release. "Hopefully, the further protection offered by the increased threatened status to endangered will help koalas survive into the future." Deakin University professor Euan

Ritchie's reaction was more morose than Crowther's and further aligned with how I interpreted the news. Ritchie called the change in the koala's listing to endangered "deeply disturbing," which is an understatement in my estimation. "Their demise is emblematic of broader federal and state government failures to sufficiently invest in the conservation of Australia's globally unique biodiversity."[1]

It's hard to believe that, despite all that the conservation community has achieved and all the resources at the disposal of government conservation programs, it's still possible, in the year 2022, for one of the world's most famous species to face the prospect of extinction. But that's what's happening in Australia today. And note how the experts didn't pin all the blame on global warming. Rising temperatures and scorching wildfires certainly are taking their toll. But worse for the fate of the koala, according to the academics who track these things, Australians and the Australian government essentially have been negligent in their stewardship of this remarkable and utterly harmless marsupial, the only one of its kind in the world. It's a sad commentary about the state of the world and the ongoing biodiversity crises: the sixth mass extinction is underway, and reporters obsessed with politics and the culture wars can't be bothered to draw more attention to it. Australians have stood by while their beloved koala is being laid low by habitat loss, invasive species, general conservation mismanagement or neglect, and now climate change, to boot (and Australia is one of the world's largest exporters of fossil fuels).

But the listing itself is somewhat hopeful. As I mentioned at the beginning of this book, what separates us from our ancestors in terms of extinction and biodiversity loss is that we modern humans care. Someone was interested in the fate of the koala, someone influential enough to get the government and Australia's scientific community to the sound the alarm. With the listing, by Australian law the government must come up with a precise recovery plan replete with timelines, anticipated budget requirements, and a list of specific actions to be taken. At least, that's what it says in Australia's foundational endangered species law.

This plan must be made public quickly, as well, and then put into action, with responsibilities expected from both federal and state wildlife management authorities. At least, that's what's expected of the folks in charge. We can only wait and see and perhaps give Australia's government the benefit of the doubt. Things didn't exactly go according to script when the world's koala population was declared "vulnerable" in 2012 following a formal senate inquiry, as University of Queensland researcher Dr. Christine Adams-Hosking mournfully recalled. "Since then, nothing has happened to facilitate the protection of koalas," she complained. "Until these root causes of the demise of this unique and ancient species are addressed with genuine conviction and real action, I cannot feel excited."[2]

The koala's listing seems to have garnered a lot of international attention. Other extinctions continue to roll along rather quietly.

In August 2022, a team of scientists announced that the dugong almost certainly has disappeared from China's waters.[3] More likely than not, it's vanished completely from the entire South China Sea region of Southeast Asia. No dugongs have been seen there for more than 23 years. This species was a cousin to Florida's manatee. Now it's a shape that will never be seen again.

Endangered species protection and recovery is a mixed bag. There are raging successes to point to, like the peregrine falcon's rebound. There are also plenty of stories of perpetually shaky and precarious situations—the California condor comes to mind, but there are other such examples to point to within and without the United States. And, of course, there are the inevitable disappointing setbacks or new concerns, as we now see with the cuddly koala. Most concerning are the multiple and mounting cases of looming extinction disasters, like the seemingly doomed vaquita of Mexico's Sea of Cortez. Many species will be lost before this book makes it to print. Most of these extinction events we won't read about or ever even know about.

A good place to view this yo-yoing of biodiversity threats and losses and hopes and frustrations is the website maintained by the International Union for Conservation of Nature (IUCN), where you'll find press releases put out by biologists and wildlife defenders hoping either to sound an alarm or to highlight where ground has been recovered in the unending struggle to protect Earth's biodiversity. The IUCN does good work, and its famous Red List and the reports it routinely issues are the gold standard in terms of monitoring the health of the planet's plant and animal life. As noted above, biodiversity conservation is a mixed bag, but the trend is generally one of decline: species are vanishing at an alarming rate, and as soon as we think one has been saved from the brink of eternal oblivion, another species gets added to the list of concerns shortly afterward.

No one would think dragonflies are at any risk of extinction. I certainly wouldn't have guessed so. But in a December 9, 2021, notice from its headquarters in Gland, Switzerland, the IUCN reported precisely that. "The destruction of wetlands is driving the decline of dragonflies worldwide,"

IUCN researchers warned in that release. "Their decline is symptomatic of the widespread loss of the marshes, swamps, and free-flowing rivers they breed in, mostly driven by the expansion of unsustainable agriculture and urbanization around the world," IUCN added in its press release. The IUCN decided to highlight the plight of dragonflies because the addition of dragonfly species to the Red List helped push the number of listed species threatened with extinction north of 40,000 for the first time in the history of the organization. Not many news outlets covered this.

The end of 2021 was a notable milestone for biodiversity but not in any good way. It seems that in much of the rapidly developing world, planners view wetlands today just as earlier planners did in the United States in the 1800s and early 1900s—as pestilent swamps needing ditching and draining in order to make way for either crops or pavement. The IUCN said wetland destruction in South and Southeast Asia now means 16 percent of the world's dragonfly species may go extinct. Fewer dragonflies means more mosquitos, I fear. There's plenty more such bad news to be found on the IUCN's website. For instance, experts predict that all coral reefs in the western bounds of the Indian Ocean may experience ecosystem collapse in 50 years' time.[4]

But there's also the occasional good news story to report.

For example, several species of tuna have seen their populations decimated for decades by rampant overfishing as Japan's sushi craze swept the world. Regional fisheries management organizations, or RFMOs, have been trying to get a handle on the disaster since the 1990s with little luck. But that finally might be changing. After years of failed attempts, IUCN now says there are signs that some tuna populations are recovering thanks to better fisheries management. Tuna species aren't out of the woods completely as fishing pressure continues, but on September 4, 2021, IUCN said that the Atlantic bluefin tuna shot up from "endangered" status to "least concern," which is, of course, a very encouraging development. Meanwhile, the southern bluefin tuna was moved from "critically endangered" to just "endangered," and the yellowfin tuna from "near threatened" to "least concern" in a new assessment. Obviously, more needs to be done to save the southern bluefin, Pacific bluefin, and other tuna species, but kudos to the responsible RFMOs that finally managed to make progress on some of these fisheries.

But this yo-yoing of species decline and rebound seems set to continue for some time. At least the overall picture has improved bit by bit compared to the bad old days of the past, especially in the United States. This wasn't always the case—not all that long ago, species' populations only declined and declined, and not one of them ever recovered.

~

William Temple Hornaday was the first director of the Bronx Zoo and an early champion of wildlife conservation in the United States. Hornaday also apparently was a bit of a racist person, according to the biographies about him that I've briefly read. Most people were like this in the late 1800s to early 1900s, so I'm not too surprised here. Unfortunate but true. I mention this only because these days too many folks construe referencing or otherwise quoting any past historical figure or presently living individual as an endorsement of all of the views held or actions taken by said individual or historical figure, especially the bad ones. So, to needlessly clarify, I do not endorse nor approve of all of the views and actions of Mr. Hornaday, who's now long deceased (he died in 1937). For those of you who understand this and can therefore tolerate brief exposure to the views of someone who may be or have been a reprehensible person in some sense—the rational and reasonable among you, that is—then my apologies for treating you like children here. For the other readers out there—the child-minded, that is—please try not to get too upset as you read on. That's because I'm about to reference and quote Mr. Hornaday quite a bit, but only because he's credited with being among the first figures in American history to raise a loud and public alarm over the extinction crisis that was roiling the whole of the United States during his days.

Hornaday's 1913 book *Our Vanishing Wildlife: Its Extermination and Preservation* may have been the very thing that ultimately saved the whooping crane and countless other species from total oblivion. Reading a few chapters of this book provides a very clear impression of just how badly the rapid expansion and development of America's population and interior in the 19th and 20th centuries ravaged the continent's biodiversity, once among the richest on Earth. Hunters were to blame, but they were inspired mainly by thriving demand for fur, feathers, and bushmeat to feed the millions flocking to America's shores and urban centers at the time. The expansion of cities, the ditching and draining of wetlands, and the cultivation of almost the entire Midwest drove earlier threats to species' survival. Hornaday clearly zeroes in on hunting.

Hornaday gives us a glimpse of the scale of the slaughter early in his book by focusing on the official harvested game tally taken in just one state, Louisiana, in just one year, 1909 to 1910. He zeroes in on Louisiana because he considered it among the states still richest in wild birds and mammals, as compared to the Midwest, where farmers had basically transformed the Great Plains into a biodiversity desert by the time his book was written.

Louisiana, by comparison, still enjoyed an abundance of wildlife, especially wetland birds, as "her reedy shores, her vast marshes, her long coastline and abundance of food furnish what should be not only a haven but a heaven for ducks and geese."[5] But he warned his readers not to expect this biodiversity richness to last for long. "The big interests outside the state send their agents into the best game districts, often bringing in their own force of shooters. They comb out the game in enormous quantities, without leaving to the people of Louisiana any decent and fair quid-pro-quo for having despoiled them of their game and shipped a vast annual product outside, to create wealth elsewhere." For the 1909–1910 season, he reports that Louisiana's state game commission recorded that hunters killed more than 5.7 million wild birds, including almost 3.2 million ducks, 1.14 million quail, and 310,000 doves, in the state. The hunters, mostly coming from out of state, also killed nearly 2 million fur-bearing mammals in Louisiana during that one year alone. And that's just what the authorities back then knew of. "Of the thousands of slaughtered robins, it would seem that no records exist," Hornaday wrote. "It is to be understood that the annual slaughter of wildlife in Louisiana never before reached such a pitch as now."

He argued that eventually Louisiana would have become barren of wild animals entirely if the state government hadn't started addressing the problem as it did, beginning in 1912 with a ban on hunting by nonresidents and stiff restrictions on the state's urban bushmeat market. Hornaday predicted that stamping out the market for wild meat likely would have the greatest beneficial impact. "A very limited amount of game may be sold and served as food in public places, but the restrictions placed upon this traffic are so effective that they will vastly reduce the annual slaughter," he said.[6]

He, of course, mentions the destruction of the passenger pigeon—the very last one died in the Cincinnati Zoo shortly after Hornaday's book was published. As I explained earlier, the passenger pigeon once flew in flocks so massive that they could blot out the sun for hours. This species was totally annihilated for the wild meat trade with surprising speed and efficiency. *Our Vanishing Wildlife* explains how bad it really was. "In 1869, from the town of Hartford, Michigan, three [rail]car loads of dead pigeons were shipped to market each day for forty days, making a total of 11,880,000 birds," he tells us. Nearly 12 million passenger pigeons killed and shipped east in less than two months, and from just one state. No wonder it took just a few decades to wipe this once remarkable species off the map completely.

The hunters who were around more than 100 years ago were very fond of killing cranes, as well. This includes sandhill cranes. Hornaday said the sandhill crane was fast on the road to extinction in his day.

For people who know this species, this may seem impossible to believe. The sandhill crane is now an abundant species easily and commonly encountered in much of the Midwest and central Canada. It's presently thriving in North America, numbering in the hundreds of thousands of individuals. Satisfied that they've failed to drive this species to oblivion, the hunters in our days want another crack at it—there is serious discussion of reauthorizing sandhill crane hunts in some Midwestern states, causing fears that whooping cranes will be killed by mistake since hunters routinely confuse the two (though they are quite different in appearance).

The sandhill crane, a species of crane so abundant as to be all but ubiquitous throughout much of North America today, was once on the brink of nonexistence just 100 years ago, as Hornaday explains in his book. He reports results of an informal survey that showed sandhill cranes had been completely extirpated from Maryland ("no record of sandhill crane for the last 35 years," Talbott Denmead of Baltimore told him), Massachusetts, Michigan, New Jersey, New York, and even Louisiana (still a wetland bird heaven back then) by 1913. They had apparently been eliminated from Wisconsin, Wyoming, and Alberta, too. Also from Minnesota, where game warden George Wood of Hibbing offered a sobering account of the scale and speed of the great North American bird slaughter for Hornaday's book. "Where there were, a few years ago, thousands of blue herons, egrets, wood ducks, redbirds, and Baltimore orioles, all those birds are almost extinct in this state," Wood wrote. On the sandhill crane, "I have not seen one in three years," he reported.

In modern times, the sandhill crane of North America is doing very well as a species. They number in the hundreds of thousands today, as I said. Its cousin, the whooping crane, isn't faring nearly as well, however.

More than 100 years ago, William Hornaday considered the whooping crane's extinction to be inevitable. He estimated that less than half a dozen existed in captivity, and wild whooping cranes were encountered so rarely back then as to be considered almost legend—like trying to find unicorns. The last confirmed sightings in the wild Hornaday points to had occurred in Canada several years before his book was published. He writes as if assuming the whooping crane had already vanished in the wild, with the species simply waiting out its inevitable oblivion in zoos, like the passenger pigeon. "This splendid bird will almost certainly be the next North American species to be totally exterminated," Hornaday said of the whooping crane, mourning its fate. "It is the only new world rival of the numerous large and showy cranes of the old world; for the sandhill crane is not in the same class as the white, black, and blue giants of Asia. We will part from our stately

Grus americanus with profound sorrow, for on this continent we ne'er shall see his like again."[7]

The whooping crane did survive, as it turns out. But just barely. It was much the same story for its old-world rival struggling to survive on the other side of the Pacific, in one corner of Asia, though Hornaday likely wasn't aware of this.

Once upon a time, in a quiet rural area of Japan, an impoverished old man was walking through a field, when suddenly he came upon a beautiful *tan-cho tsuru*, a red-crowned crane. The crane was injured—she had been shot by careless hunters and left there to die. Appalled, the old man scooped the beautiful, wounded crane into his arms and quickly carried her to his residence.

The old man was poor and didn't have much to offer the crane, but he rushed her home still. At his home, the old man carefully laid the crane down to bed and dressed her wounds. He fed the crane and gave her water to drink. Day by day, the crane gradually recovered and regained her strength. Eventually, the man successfully restored the crane to full health. When the time came for them to part ways, the crane bowed gracefully to the old man, showing her deep gratitude, then walked out the door and flew away into the sky. The man smiled and returned to his impoverished life.

A few days later, the man heard a knock at his front door. He opened the door, and there stood a beautiful woman. "I have nowhere to go," the beautiful woman said to him. "Please allow me to stay with you, and I will be your wife," she pleaded with him. "But I have nothing," the old man protested. "You are beautiful and I would be honored to take care of you, but I cannot. I barely have enough for my own needs. I am very poor." He pleaded with her to find someone else. The woman pressed on. "Do not worry. I have this bag of rice," she showed him. "It is enough to feed the both of us." The man shook his head. "But only for one night. I'm sorry, but once the rice is gone, we will starve." The woman persisted. "Do not fear. This bag of rice will never go empty. Trust in this. I swear it." The old man saw that the woman was in need and would not take "no" for an answer. He relented and reluctantly invited her in.

It was as the woman told him. That evening they cooked the rice and ate until they were both full. The next day, the rice bag was full again. And again the next day. And the next day. And the next. As she had promised, the bag of rice never went empty, and the two were always well fed.

The days passed, and the woman was good to the man. She cooked for him and fed him and took great care of him. He no longer had to fear hunger. But still, they were very poor.

The woman then said to the man, "I will go into the other room and I will make a beautiful kimono. This, you will sell in town for a good price, and our poverty will be a thing of the past. However, you must promise to never enter the room while I am working. You must leave me alone to work in peace. Promise me this." The man swore it. And so the woman walked into the next room and closed the door.

For several days she sat there, working away in solitude, and the man left her alone and never once entered the room, as she had insisted and as he had promised. After about a week, the woman exited the room and presented to him the most beautiful and luxurious kimono he had ever seen in his life. "Here, it is finished," she told him. "Take this to the market in town. You will sell it for a very good price. And then, we will no longer be poor." The man did as he was told. The shop owner in town was deeply impressed with the beauty and superb quality of the kimono. He practically begged the old man to sell it to him. The man did so, as the beautiful woman told him to, and at a very good price. The man returned home to the woman with a sack full of money. They were no longer poor.

Days passed by, and then the woman said again to the old man, "I will go back into the room and will make another kimono. This, you will sell to the market again, as before. With this second kimono, I believe you will get such a good price that neither of us will ever have to work again. We will be free from poverty for good, for life." The woman walked into the next room and closed the door, and the man once again promised to never open the door and to leave her be until she had finished her work.

About a week passed as she toiled away. More days passed by, and the woman continued her work. The man grew impatient, then curious. Eventually, he couldn't take it anymore, and he let his curiosity get the better of him. He walked to the door.

Very quietly and slightly, making almost no sound, he opened the sliding door, just a slight crack. He peeked inside and was astonished by what he saw.

There, at a large loom, sat not the beautiful woman but a red-crowned crane, the very one the old man had saved in the field weeks earlier. The crane was removing its own feathers and using these to make the kimono.

The crane turned its head and saw that the old man had broken his promise. "You saved me those many days ago, and I am deeply grateful," the crane told him. "This is how I sought to repay your kindness. But you have broken your vow. I am sorry, but now I must leave." In what seemed like an instant, in a gust of wind the crane then flew out the window, leaving her freshly plucked feathers and the unfinished kimono resting on the floor.

The old man never saw the woman or the crane ever again, but he understood. The beautiful woman, the one who fed him and saved him from poverty, was the red-crowned crane all along. She had come to repay him, to demonstrate her gratitude to him by ending his life of crushing poverty. But now, she was gone.

The end.

~

That's my rough, Cliffs Notes version of a very old and classical Japanese folktale. The Japanese know this very famous story as *Tsuru No Ongaeshi*. One popular English translation of this is "The Grateful Crane." A more literal translation would be something like "the crane's returned favor," but that, of course, sounds a lot better in the original Japanese. I'm not quite sure how old this story is, but it dates back at least several centuries, possibly as far back as the Heian period or even earlier. There are several variations to this tale, but they all have more or less the same plot: a man saves a crane, and the crane returns magically transformed as a woman to show her gratitude by helping him in some significant way. But the man discovers the crane's true identity, and then she flies away, and the man never sees her again. The tale is a play on several folk beliefs popularly held in Japan in the past. One Japanese legend holds that a red-crowned crane can live for 1,000 years, and should a person happen upon a crane in distress in some way and help it, then that crane would grant that person a wish. The Ainu, an indigenous population that inhabited Hokkaido before the Japanese arrived, called red-crowned cranes *sarorun kamuy*, or "marsh gods," and held them in similar reverence.

You know origami, right? The Japanese art of paper folding. You almost certainly know of the famous paper crane that is probably one of the very the first things that students of origami learn how to make, that angular figure with the pointy beak that sometimes flaps its wings when you yank on its tail, depending on how it's folded. *Tancho* is that crane.

The red-crowned crane is very famous and known to all Japanese. I've never met a Japanese person who didn't know this species. Before moving to Texas those many years ago, I had barely even heard of the whooping crane, and I was working as an environmental journalist at the time. The comparative cultural statuses of these two species are quite different. I am unaware whether the whooping crane was held in similar special regard by Native Americans and made the subject of similar oral legends, but there is no doubt that the red-crowned crane was and is far more culturally significant to the Japanese than the whooping crane is to us modern Americans.

We know that the red-crowned crane once inhabited much of the Japanese archipelago, nearly anywhere on Hokkaido, where its preferred habitat can be found, and a wide swath on the main island of Honshu for sure. This graceful wetland bird appears again and again in Japanese classical art. I've seen images of red-crowned cranes painted on the walls of 600-year-old shrines in Kyoto, which is about 1,000 kilometers from the cranes' present home base in the Kushiro region. Heck, go to any museum in the United States that houses a Japanese classical art exhibit, and your chances of spying an image of the red-crowned crane somewhere are very good. But as Japan's population swelled and the country expanded deeper into the woods and wetlands, the crane's numbers fell and fell. As in America, the ditching and draining of wetlands for agriculture took a great toll, as did relentless and uncontrolled hunting, especially after bows and arrows (which likely wounded the grateful crane of that legendary story) gave way to firearms. Hornaday would have recognized the story.

It's hard for me to get my head around this, but even culturally significant species can be driven to the brink of extinction (think of the panda in China) or to total oblivion (like the moa in New Zealand) by simple human greed. Animals of legend were once hunted relentlessly, the same as animals held in no particular esteem or regard. By the late 1800s, the Japanese assumed that the same basic forces that had ended the moa and dodo had also sealed the fate of their beloved red-crowned crane. The government and nearly everyone else in the country were of the opinion—a very reasonable one to hold at the time—that all living red-crowned cranes had been extirpated from Japan completely.

Wasn't it Joni Mitchell who said something like you never seem to know what you already have until you've lost it? The Japanese felt the same way. But it was too late to bring back this legendary bird of Japanese folklore.

Or so they thought.

It was around this time when the Japanese overthrew the Bakufu military dictatorship and embarked on the Meiji Restoration, a period of rapid development and modernization. Japan's new government grew determined to see Hokkaido settled and developed, as well—Hokkaido was Japan's last frontier, something akin to Alaska for the United States, except that Alaska was sold to the United States by a disinterested Russian Empire, which very much had its eyes on Hokkaido back then.* Fearing further Russian encroachment, Japan set out to make Hokkaido fully Japanese and launched the great *kai-taku*, or taming of the north island. That saw the government sending expeditions and survey crews with the aim of exploring and gathering as much information about the island as possible.

By the early 1900s, the extinction of the red-crowned crane from Japan was an established fact. Then one day, as the story goes, sometime in the 1920s the Japanese government dispatched a survey team to explore some remote corners of East Hokkaido. They were surveying the expansive Kushiro Marsh wetland there, Japan's largest, north of the port city of Kushiro, when they suddenly encountered, to their amazement, a small but very much surviving population of red-crowned cranes taking advantage of the hot springs and the marsh's relative abundance of forage. It's said that the discoverers initially couldn't believe their eyes. Imagine if the Louis and Clark expedition had stumbled on a surviving herd of woolly mammoths or a wayward family of saber-toothed tigers during their survey of the Missouri River headwaters. The effect was similar in Japan.

Word of the crane's surprising survival quickly spread, and the government just as quickly mobilized to save the species. Hunting was permanently banned, and the area of Kushiro Marsh was set apart under government control and out of the hands of the encroaching farms and cities of the region (it was too late to save the coastal portion of the marsh, where Kushiro is built). Japan's attention to *tancho* later waned during the war but revived shortly afterward, despite the deep poverty in Hokkaido during those years.

In 1950, the community of Akan began feeding its resident red-crowned cranes during the winter months after growing concerned that the population numbers weren't recovering all that well. Shortly afterward, a dairy farmer named Itoh Yoshitaka, living near the young village of Tsurui (the village's name loosely translates to "where the cranes live"), did the same. Mr. Itoh set aside about 10 hectares of his land and established a feeding station for the cranes, today called the Tsurui-Itoh Tancho Sanctuary, bringing the number of winter feeding stations in the region to two. A third winter feeding station was later established just south of Tsurui, at a point off the highway called Tsurumidai (basically "crane viewing station"). These feeding grounds were originally operated entirely by the community, but central government conservation authorities later assumed control of them and

* I read once that near the end of World War II, the Soviet Union asked the United States to let it occupy the northern and eastern half of Hokkaido, establishing Kushiro as its base of operations in occupied Japan. The Americans apparently threatened to turn their guns on the Russians if they tried this, and Moscow backed off.

folded them into Japan's formal red-crowned crane population rehabilitation strategy. They are now among the nation's most popular tourist attractions.

I first encountered the red-crowned crane on a drive south of Akan, where the community-driven effort to save the cranes began in the 1950s. My wife and I rented a car in Sapporo, and we made the nearly four-hour drive east to Kushiro to explore and look at houses or land, as we were interested in settling down in the area at the time. We desired a more rural setting, and so we set out exploring the back roads between the main area of Kushiro and the village of Akan.

I turned off the highway and onto a side route past farms and groves of trees. I was glancing to the left at an open field at one moment when my wife suddenly shouted to me, "Look out!" I pressed down on the brakes and then scanned back and forth, expecting to see a deer about to jump in front of our rental car. What I saw instead took my breath away.

A large, adult red-crowned crane, standing close to five feet in height, was quickly making its way from a ditch, strutting proudly and advancing across the road as our vehicle was approaching. The crane seemed determined—hell-bent even—to continue its mission to cross the street, cars be damned, so I would've hit it or come very close to hitting it had my wife not alerted me to the danger.

I quickly brought our vehicle to a halt only about 20 feet or so from where the crane stood. It never stopped walking forward. The crane simply glanced at us with disdain, briefly, and then stubbornly made its way onto the road right in front of us, pretty much ignoring us. Luckily, no other vehicles were behind us at the time. I switched on the vehicle's hazard lights just in case. And there we sat, parked in the middle of the road, regarding this amazing creature while waiting patiently for it to cross to the other side. It took its sweet time, of course.

Two more cars soon came from the opposite direction, but by this time the crane was plainly visible in the middle of the street. Those cars both came to a stop, as well. And there we all waited patiently. The crane ignored us completely and casually made its way across the road, head held high as if it owned the place. In a way, it kind of did. The red-crowned crane is a celebrity in these parts, after all. "Why doesn't it just fly across?" I finally asked. My wife didn't have a good answer. Probably because it didn't have to. In East Hokkaido, traffic usually yields to the cranes, not the other way around.

I would later encounter such a scene on multiple occasions while enjoying my life in East Hokkaido. I experienced many, many other close encounters with the red-crowned crane in and near Kushiro, Akan, and Tsurui, a town whose entire identity is shaped by the crane (Tsurui's town mascot is Tsurubo, a lovable and near perfectly cylindrical rendering of a cartoon *tancho*). I may have struggled to get within decent viewing distance of the antisocial crane back in southeast Texas, but the grateful crane is not a shy bird by any stretch of the imagination.

They flew from the Itoh sanctuary just a few feet over our heads, casting long shadows like jumbo jets approaching an airport runway. They are spotted easily on drives through the countryside and make themselves available for closeups with photographers in Akan and Tsurui every winter. If you happen to stumble upon a crane nest in the marsh during the spring, they almost certainly will attack you; they won't fly away in a heady panic like a whooper might. Though the Japanese almost eradicated them, the red-crowned cranes have a reputation for being tough birds. They're known to fight off large birds of prey like the Steller's sea eagle, one of the world's largest eagles and an endangered species, and other predators in defense of their young. This behavioral trait may have facilitated their more rapid recovery compared to the whooping crane, but from my research, that's not likely. Rather, it seems abundantly clear to me that the main factor behind Japan's great success with red-crowned crane rehabilitation and population recovery is the winter feeding program launched independently by villagers in the 1950s.

Now for a little bit of a more academic overview of this remarkable species and its beautiful home in East Hokkaido.

Kushiro Marsh National Park is a protected zone encompassing some 28,800 hectares located in the southeastern corner of the island of Hokkaido. It's registered under the United Nations Ramsar Convention on Wetlands of International Importance—signatories were directed to designate at least one major protected wetland as the price of participation in that treaty body, and Japan specifically chose Kushiro Marsh as its first signature national wetland when it signed and ratified the Ramsar agreement. Today, Kushiro Marsh National Park contains unique habitat for more than 2,000 species of wildlife.[8] But its most famous and iconic resident is, of course, the red-crowned crane, *tancho*, a large wetland bird famous for the elaborate and often beautiful winter courtship dances that take place in the snowy backdrops of eastern Hokkaido.

Though pristine and relatively undisturbed throughout much of its expanse, Kushiro Marsh National Park is surrounded on almost all sides by urban and agricultural development. Though the park itself has remained mostly protected and relatively undisturbed, the area surrounding the park has experienced massive land use changes since those pioneers surveyed the interior north of Kushiro more than 100 years ago.

In the south, most of the original coastal wetland habitat has been overtaken by the city of Kushiro, as noted earlier, which was built atop the southernmost coastal wetlands to develop a continually ice-free port for fishing (likely the feature most appealing to the Russians). Thus, the construction of Kushiro cut off most of Kushiro Marsh from the sea. To the west, north, and east, the marsh is boxed in by agriculture, mainly dairy farming operations and fields for raising hay used to feed dairy cattle. The region is famous for dairy, and the dairy industry there is mostly thriving, in sharp contrast to the situation in the United States, in which dairy farmers are struggling. Research on the potential for abandoned farmland in the region surrounding Kushiro Marsh to become additional habitat for the red-crowned crane is ongoing, though initial results suggest these lands immediately adjacent to the marsh, seemingly ideal candidates for wetland expansion (since they are no longer under cultivation of any sort), are not expansive enough to make up for historical wetland losses.[9]

Although the marsh itself has long been protected (registered under the Ramsar Convention in 1980, later established as a national park in 1987), extensive redevelopment has occurred on the waterways in the north beyond park boundaries without regard to the potential negative impacts on downstream ecosystems. The biggest land use impact occurred following work to straighten the rivers and streams lying north of Kushiro Marsh. For example, work was performed in the 1970s and 1980s to re-channel and straighten portions of the Kuchoro River, a major East Hokkaido waterway that drains into Kushiro Marsh. The work to straighten the Kuchoro River was done in an attempt to prevent upland flooding damage to the dairy operations.[10] Of the 45 kilometers of the river's main channel length, 10 kilometers of the river just outside the park's boundaries were eventually straightened.[11] That's a little over six miles.

This is an old story in the U.S., as well. Early developers looked at a naturally winding river and said, "How inefficient is that?" They figured frequent floods occurred because rivers and streams didn't have a straight shot to the sea to move the water out of the area as quickly as possible. In reality, rivers wind naturally through terrain, especially flat terrain, like where the Mississippi River sidewinds its way through the plains of the Midwest and south into the Gulf of Mexico, because that *is* the most efficient way for a river to run. Nature knows what it's doing—the river carries not only water, but also sediment, and a winding, squiggly path is the natural and most efficient way for a river to move both over considerable distances. But for decades "experts" both in Japan and the United States didn't fully understand this. Now they do, and work is underway in both countries and elsewhere to "rewind" previously straightened rivers in a bid to restore ecosystem function and watershed health.

Several studies conducted since the straightening of upland rivers in East Hokkaido have confirmed that this work resulted in a major influx of sediment into Kushiro Marsh, effectively "drying" significant portions of the wetland.[12] The straightening of other upland rivers and streams, especially the Kuchoro River, resulted in a 0.1 to 0.2 percent steeper gradient to the river system, thereby increasing the energy of the river.[13] They made the river straighter, but also the elevation of its descent steeper, causing it to run faster and stronger. Researchers Nakamura and his colleagues found that during the past two decades this steeper, faster run of the Kuchoro River resulted in the river cutting itself deeper and deeper into its own riverbed, ultimately carrying and depositing greater volumes of sediment further downstream until the waters reached the lower elevations at Kushiro Marsh. The waterways in the marsh itself were never straightened and were in fact protected from such engineering by the designation of the entire area as a protected area and national park. So when the river system flooded in the marsh itself, it deposited this extra sediment load when the floodwaters receded. The river straightening work added soil to Kushiro Marsh.

What does this eventually end up doing? As it turns out, additional sedimentation provides a more stable soil environment for hardwood plant species to take hold. Wetland species are pushed out, and as a result the area of wetland declined gradually as more hardwood-type species took over. Hardwood habitat expanded, and wetland habitat declined. This made Kushiro Marsh, in effect, drier. Dendrogeomorphological sampling and radionuclide testing of sediment deposits in the marsh have linked the added drier soil layers to specific major flooding events in Kushiro Marsh National Park that occurred shortly after work to straighten portions of the Kuchoro River was completed, so scientists know the two are connected.[14] It's not just coincidence. Studies have found that within a year of the first major flooding event to happen in Kushiro Marsh following river straightening, hardwood species, especially the common native species Japanese alder, established wider root systems in this newly laid sediment and began thriving in areas where earlier it would have been impossible for these heavy hardwood trees to grow.

What does this all mean? It means that throughout much of the red-crowned crane's recent conservation history, especially during its quickest bout of population recovery, the species has been losing available foraging habitat in Kushiro Marsh. By the late 1990s, the original wetland expanse within Kushiro Marsh National Park was estimated to have declined by about 20 percent while Japanese alder forest cover increased from 8.6 percent to 36.7 percent.[15] And that was due to the work undertaken to mess with Hokkaido's natural upriver hydrology.

During the past 100 years, Hokkaido has lost some 70 percent of its original wetland area to agriculture or urban development. Most of this degradation of the natural environment came in the form of direct conversion of wetlands to farmlands.[16] The situation in Kushiro Marsh has since apparently stabilized, thanks in part to intervention efforts. Tree cutting and strand removal of Japanese alder by Kushiro Marsh National Park managers have been found to be effective means of both protecting the surviving wetland areas and promoting the restoration and expansion of wetland vegetation. A 2012 field experiment found that Japanese alder removal encouraged the recovery of natural wetland hydrodynamics, which then promoted the growth of more common native wetland vegetation such as grasses and reeds more favorable to the red-crowned crane and other iconic Kushiro Marsh species.[17] Other attempts at mitigation or ecosystem restoration efforts in East Hokkaido include engineering projects aimed at "rewinding" portions of artificially straightened rivers in the higher elevations to prevent further downstream sediment loading. But in many ways the damage is done.

However, investigators have turned their attention to the potential for abandoned farmlands in the region to eventually become incorporated into existing wetland expanses to offset habitat lost to the Japanese alder intrusion.

For instance, researchers have delved into the question of whether wetland animals are beginning to colonize abandoned farmland adjacent to Kushiro Marsh, beginning with the lowest order species in the food chain. A study led by Yamanaka and colleagues focused on the propagation of wetland ground beetles in abandoned farming regions, setting out to determine whether occurrences of these species in abandoned farms represented occasional wanderings from the beetles' home habitat in Kushiro Marsh or if these beetles were indeed establishing themselves permanently in these newly available spaces no longer under the plow.[18] The answer seems to be the latter—the researchers concluded that the presence of wetland ground beetles in abandoned farms in and near Kushiro Marsh represents permanent establishment and colonization, thus an expansion of the beetles' habitat in the region. That study found that an increase in the soil moisture content in these lands left behind when the farms shut down and agriculture ceased results in ecosystem conditions more closely resembling the beetles' preferred Kushiro Marsh wetland habitat, raising hopes that the resident red-crowned cranes may one day find these tracts appealing as well, especially for breeding. As *tancho*'s numbers swell, it will need far more room to spread its wings.

Red-crowned cranes are known to exist in two main population concentrations: the flocks that migrate from northeastern China and parts of the Korean Peninsula to grounds in eastern Siberia and the nonmigrating flock located in eastern Hokkaido, Japan, the focus of my research. The population in Hokkaido has expanded rapidly, while populations in China and Siberia are believed to be in decline.[19] Hokkaido's crane population is now believed to be larger than mainland Asia's. Conservation efforts targeting red-crowned cranes on the Asian mainland are less well defined and developed than in Japan and have a shorter history. For instance, a recent study out of South Korea noted that a local government only recently became interested in advancing ecotourism centered on cranes but had no plans to protect or manage the habitat, thus likely harming the wintering population there.[20] In parts of China and Russia, migrating flocks have suffered population declines of 50 percent or more due to hunting, poisoning, and habitat loss.[21] By contrast, the red-crowned crane population on Hokkaido has shown impressive resilience and consistently strong population growth over a 70-year history, despite the fact that the Japanese red-crowned crane in Japan was nearly wiped out entirely due to habitat loss and overhunting, as noted earlier, the same pressures now pushing red-crowned crane numbers lower on the Asian mainland.

Historical accounts by the Red-Crowned Crane Conservancy (RCCC) and Japan's Ministry of the Environment state that in 1924 a Japanese government survey team discovered a small flock of some 20 red-crowned cranes surviving in Kushiro Marsh in eastern Hokkaido. Other survivors were soon discovered in smaller numbers, and the estimated number of surviving cranes rose steadily over the subsequent decades (RCCC data shows 33 red-crowned cranes in Hokkaido in 1952).

The national government originally set out to protect this surviving population through stricter enforcement of a hunting ban, but by the 1950s Japanese conservationists feared the crane population was still recovering far too slowly for comfort. Local communities took it upon themselves to save *tancho*, and thus the seasonal winter artificial feeding operations were launched and later granted government sanction, the first beginning at the present site of the Akan International Crane Center in 1950. Two more stations were later added in the village of Tsurui, as noted. These are all considered very popular roadside tourist attractions today.

There was an immediate positive impact to this endangered species. The earliest available data shows the red-crowned crane population expanded exponentially in the early decades following the launch of the organized winter artificial feeding program.[22] By 2008, the East Hokkaido red-crowned crane population was estimated to be about 1,241 birds, this despite a harsh winter experienced that year, according to data compiled by the Tancho Conservation Research Group. RCCC data put the 2014 population at more than 1,400, and its most recent figures gathered in annual winter census counts suggest that there are around 1,800 to 1,900 red-crowned cranes living in Hokkaido today, up from just 33 in 1952. The population has rebounded quite well, though it's taken quite a long time for it to do so.

As I mentioned, the red-crowned crane looms large in Japanese culture, art, folklore, and literature. Today, the red-crowned crane is a celebrated symbol of eastern Hokkaido and one of the region's top tourist attractions, especially famous for its winter courtship dance adored by photographers the world over. The main regional airport in Kushiro is named for the crane (Tancho Kushiro Airport; *tancho* meaning "red crown"), and the regional ice hockey team is called the East Hokkaido Cranes. The three feeding stations at Akan, Tsurui, and Tsurumidai are top draws for nature photographers, while tourists also flock to see cranes at rehabilitation centers at Kushiro Zoo and the Red-Crowned Crane Natural Park (not counting, of course, the time Japan decided to completely close itself off to foreign tourism in an attempt to contain the COVID-19 pandemic). Red-crowned-crane-themed souvenirs are popular with millions of visitors passing through the region every year. The red-crowned crane's historical, cultural, and economic significance for eastern Hokkaido cannot be overlooked in any discussion of this species' conservation future.

Since its establishment, the winter artificial feeding program has been the prime vehicle for red-crowned crane population management and recovery policy in Japan. Habitat conservation has played a key role as well, of course, in particular the designation of Kushiro Marsh as a federally protected national park in 1987. But because conservation managers have little control over land management decisions outside the park and due to constraints on expanding the park's boundaries, artificial feeding and direct population management have been the focus of red-crowned crane conservation initiatives for decades. Unfortunately, the Ministry of the Environment has its writ limited to just the national park's boundaries and the feeding stations, which are located outside Kushiro Marsh National Park land. Efforts to "rewind" the Kuchoro River and other upland waterways in an attempt to protect Kushiro Marsh from further excessive sediment loading are overseen by the Ministry of Land, Infrastructure, Transport, and Tourism.

The feeding operations at the three winter feeding stations in East Hokkaido are simple and straightforward affairs. The work consists of grinding corncobs and then scattering the kernels atop the snow and ice at the cranes' feet. The first time I saw feeding underway happened during a site visit to the Tsurui-Itoh Tancho Crane Sanctuary. Feeding occurred at approximately 1:00 p.m., whereby a single sanctuary official hauling a big cart filled with feed used a large scoop to scatter the corn at the feet of the visiting cranes. The official moved quickly through the sanctuary and scattered the feed as widely as possible until his cart was empty.

It was at this time that I experienced just how thoroughly habituated the red-crowned cranes are to the human presence and this winter feeding routine. A few of the cranes waited immediately outside the door to the sanctuary storehouse in anticipation of the feeding. They knew food was coming and at approximately what time. Once the corn was scattered, the lone sanctuary official hustled back indoors and left the wild cranes unsupervised, free to gorge themselves as they wished. The cranes ate their rations greedily and paid no attention to the 50 or so photographers gathered from all corners of the world to document the annual spectacle. The winter artificial feeding program is managed and overseen by the Ministry of the Environment's regional office in Kushiro. MoE is now determined to end winter feeding in an effort to encourage the eastern Hokkaido population to expand its range and begin settling and breeding in other parts of Japan. Though the Kushiro and East Hokkaido regional population is decidedly nonmigratory, the Japanese red-crowned cranes have retained their natural ability to fly long distances as their cousins on the Asian mainland do. Some have flown impressive distances away from Kushiro Marsh, as later studies and news reports confirmed.

A plan to wean the red-crowned cranes off winter feeding was first finalized in early 2013, and implementation began in earnest in 2015. The initiative was reviewed by the ministry in 2018. The strategy calls for the three feeding stations in Tsurui and Akan to reduce the volume of feed disbursed to the cranes by approximately 10 percent annually, with adjustments anticipated as necessary, continuing every year until eventually winter feeding is halted entirely. Authorities agreed to periodically evaluate progress and allowed for potential adjustments to the program as deemed necessary.[23] A 2018 audit of the program found the three feeding stations to be largely in compliance with MoE's instructions. But the ministry eventually faced protests and opposition to its plans.

The volume of food disbursed to cranes at the Itoh-Tancho Crane Sanctuary fell from a record high of 5,250 kilograms to 3,960 kilograms, from 5,250 kilograms to 3,370 kilograms at the Tsurumidai station, and from 7,740 kilograms to 3,800 kilograms at the Akan International Crane Center. In total, the ministry's 2018 review found that combined winter feeding had fallen from 18,240 kilograms to 11,130 kilograms, a decline of approximately 38.9 percent.[24] The 2020–2021 winter seasonal feeding was scheduled to fall to at least 50 percent of historical high volumes according to the existing plan. Despite the steady decline in the volume of feed tossed at *tancho* each winter, as of this writing, the latest population census data issued by RCCC showed the Japanese red-crowned crane population holding steady or continuing to expand. Some point to the numbers as a worrying indication that perhaps the population has stopped growing, blaming MoE and its feeding reduction plan for this. I doubt population growth has stalled for this species, but the jury is still out.

Let's take a closer look at the red-crowned crane's main home on the beautiful island of Hokkaido.

Kushiro Marsh National Park is by far Japan's largest wetland. Located adjacent to the city of Kushiro in the southeast corner of the island of Hokkaido, Kushiro Marsh alone comprises some 60 percent of Japan's total wetland area.[25] The national park, among Japan's largest, is home to approximately 1,400 species of animals, including 170 separate bird species and some 1,150 species of insects and invertebrates. Its most famous resident is, of course, the red-crowned crane, but it's not the only frequently spotted resident. Where the ground is firm enough, you can see a hearty species of northern fox and one of the world's largest deer species, *ezo shika* in Japanese, or sika deer in English. Kushiro Marsh was designated as a wetland of international significance under the Ramsar Convention on Wetlands in 1980. Prior to the global COVID-19 pandemic, approximately 600,000 people visited Kushiro Marsh National Park every year, including thousands of visitors from abroad.[26] The park, the most accessible in the region, is popular for its summer and winter canoeing and kayaking opportunities, nature trail hiking, bird watching, and scenic vistas, and it is one of several major attractions in East Hokkaido.

Researchers have determined that during the past 50 years or so Kushiro Marsh has shrunk by some 20 percent from its original total area, representing a loss of approximately 25,000 hectares, according to an assessment by the Ministry of the Environment. The increased sedimentation mentioned earlier has been found to be the root cause of this steady contraction of wetland area. The excess sedimentation and de facto drying of the marsh means a rich community of reeds and sage grasses gradually has been giving way to woodier plant species, especially Japanese alder, in turn shrinking the area of suitable habitat for the endangered red-crowned crane.

Ministry of the Environment researchers have determined that a dramatic reduction in catchment sedimentation is the only way to halt the shrinking of the wetland. In other words, unless they can completely reverse the upriver straightening work done to the Kuchoro (and they can't rewind the entire engineered length, unfortunately), then they have to find other ways to prevent new sediment loads from filling the downriver wetland every time it floods. To that effect, the Nature Restoration Council for the Kushiro Mire was established in 2003 per a national Nature Restoration Promotion Law. The Restoration Council is comprised of six subcommittees that plan and operate various restoration projects. The council also monitors scientific research conducted in the national park and its vicinity. To date, the council has organized three main marsh restoration projects: incorporating abandoned farmlands into crane habitat; a reforestation campaign around Takkobu Lake; and the main initiative, that "river rewinding" project I mentioned earlier that aims to restore the natural riparian processes to the waterways.

The largest river rewinding effort actually involved not the Kuchoro River, but the Kushiro River, in an area north of the main marshland in the Kayanuma region, where work restored two kilometers of its natural run.[27] Some deemed the Kuchoro to be the biggest problem in terms of newly deposited sediment loads, but other studies argue that the greatest source of excess siltation stems from the unnatural channelization of the Kushiro River. Neither is the only source of the problem, as "the widespread attenuation of sediment transport and annual floods has led to the encroachment of riparian vegetation onto the former gravel beds along many rivers," as researchers Amano and colleagues put it. The upland watershed has been altered in multiple ways during the period when settlers poured into East Hokkaido. Various other rewinding and nature restoration projects have occurred or are underway "at rivers, moors and tidal flats including Kushiro Shitsugen, based on their respective Grand Designs and Implementation Plans developed by respective Nature Restoration Committees which consists of governments, experts, NPOs, local residents and others related," per a study by researchers Nakajima and others in their description of the bureaucratic mess that lay behind the management of Kushiro Marsh National Park.[28]

Environmental education also plays a major role in Japan's wetland conservation and greater restoration initiative. But river restoration remains the prime goal of the Nature Restoration Council, with most efforts aiming "to restore the meandering Kushiro River and the habitats there," according to a report by the Kushiro Nature Conservation Office.

The second most discussed strategy involves incorporating tracts of abandoned farmland into the boundaries of Kushiro Marsh National Park in the hope that a gradual return of natural hydrological functions will encourage the expansion of native marsh plants and thus crane habitat. The thinking seems to be that if the wetland is drying in the middle, then perhaps the edges could be made a bit wetter.

And here, I go off on another tangent.

There is little to no serious discussion in the existing academic literature that I could find regarding managing or actively altering abandoned farmlands adjacent to Kushiro Marsh National Park for erosion control purposes. Likewise, no serious effort exists aimed at managing any adjacent lands controlled by some very large dairy operations for the same purpose. In my opinion, this is a major gap in the discussion over this marsh's management and conservation, since the surrounding dairy lands are almost certainly eroding and adding unwanted additional sedimentation on their own. They're a missing piece to this important conservation puzzle.

Abandoned farms at lower elevations may be transformed into wetland habitat, but what about lands at higher elevations gradually eroding lower into the marsh itself? Yes, conservationists have been tracking the gradual loss of wetland area within Kushiro Marsh National Park for several years now, and the alteration of waterways in the watershed has been deemed the main culprit, as it's disrupted the natural process of erosion that has occurred for thousands of years there. The extra sediment load is leading to the gradual expansion of drier lands and the intrusion of woodier plant species to the detriment of wetland grasses and reeds, habitat favored by the red-crowned crane. Still, it seems to me that mitigation shouldn't simply involve "rewinding" the upland waterways. The area immediately surrounding Kushiro Marsh is dominated by dairy farming operations, with land used either to exercise and graze cattle in the warmer months or for the cultivation of hay as cattle feed for use in concentrated feed lots. Despite the focus on the water channel diversions as the source of the excess sedimentation of the marsh, to date no focused effort has been organized to control the siltation problem via addressing the alteration of the surrounding landscape for dairy cattle operations, at least none that I can find.

One way to correct this missing component in the region's watershed management system might be to address the siltation problem in the marsh by shifting focus to the private lands surrounding the park, the lands cultivated for dairy operations. A new erosion management plan could reduce runoff from the agricultural land to the streams and rivers by, for example, establishing a program for landowners to plant vegetated buffer zones between their farms and the main outflowing streams and rivers. Native vegetation, in particular the hardwood species Japanese alder, could be beneficial here, though Japanese alder is now considered the main scourge of the marsh interior and the most visible indicator of the shrinking of Kushiro Marsh.

Alder has been recognized as a beneficial species for thousands of years, but it also was discovered to have nitrogen-fixing properties by the 19th and 20th centuries. Research in the 1970s found that "atmospheric nitrogen is fixed symbiotically in root nodules of *Alnus*."[29] Japanese alder leaches nitrogen from roots and nodules into the soil, stimulating both additional plant species growth as well as microbial activity. The leaf litter and debris from alder trees can add nitrogen to soil via surface leaching. Because of these properties, alder has been identified as an ideal erosion control species due to its root system strength and nitrogen fixation that helps to enhance biomass in the soil and binds soils more tightly from the added microbial communities and fungi. Native Japanese alder trees already are being employed in one erosion control program occurring in the vicinity of Kushiro Marsh, the Takkobu Native Forest Restoration project. Sure, the choice of Japanese alder is ironic, since it is the species deemed most responsible for the loss of marshland from siltation and drying—the trees colonize the core of the wetland and then spread, to the detriment of wetland reed and sage. But this is still a native species, one shown to thrive in the local climate where the average annual mean temperature is just six degrees Celsius. Given alder's proven capacity to alter soil quality and composition through nitrogen fixation and erosion control, the species is probably ideal for use in anti-erosion buffer strip creation.

Again, alder has been shown to increase the soil's resilience to erosion factors by fixing nitrogen and thus encouraging greater microbial and mycorrhizal bacterial activity, which helps bind the soil but also increases the soil's water-holding capacity, thus reducing the impacts of compaction by cattle and surface and subsurface erosion via runoff into adjacent streams. You don't want these trees growing within the protected wetland itself, of course, but there's no real reason not to use them for erosion control outside the park's boundaries. That's just my two cents, anyway.

It wouldn't be too hard. Conservation authorities could start by designing and implementing an erosion control program centered on the upland area dairy industry using satellite photos to help locate optimal tracts for targeted erosion-control measures. An incentive structure could be implemented to encourage landowners to set aside tracts of their land adjacent to waterways or major runoff areas for growing alder-based natural vegetated buffer zones to encourage downstream erosion control. To avoid problems with invasive species (already an issue in the wetland), the project will select native plants with which to build these buffers, either via transportation of adults and seedlings from the marsh region where feasible or via new planting.

If one wanted to make an experiment out of this, satellite, GIS monitoring, and on-the-ground monitoring could be employed to measure erosion rates in buffer areas versus some identified control sample plots nearby where no vegetated buffers are set up. The wetland's protectors may be surprised by what they find. At any rate, what I'm arguing here is that missing from the current list of conservation and restoration projects is a focus on sediment loading from the erosion from adjacent dairy lands. It's probably a major part of the problem—hydrology experts fully understand that agricultural practices, including grazing and rearing cattle for dairy, accelerate erosion rates in existing watersheds, as the cows' hooves compress the land and kick up a lot of dust and debris at the same time.

Sorry about all that. I'll stop obsessing over Japanese alder. The main point is to demonstrate that, in endangered species and habitat management, as in life, it's quite possible and often even desirable to make an ally out of an enemy. Are invasive or intrusive species always and only enemies, or do they sometimes provide useful ecological benefits? Something to keep in mind.

End tangent/rant.

Now let's take a moment to step back to compare and contrast whooper and *tancho* side by side to see what we've learned thus far about these amazing birds.

Of the eight species of cranes in the genus *Grus*, *Grus americana* and *Grus japonensis* are the most similar to one another, close enough in both size and outer appearance to suggest that interbreeding between these species may be possible, though that's not an experiment anyone would allow me to conduct, and for good reason.

I use the whooping crane and red-crowned crane to give my students a lesson in speciation. It is plainly evident that the two are very closely related, yet one species is in Asia and the other is in the middle of North America. They're separated by half a continent and the world's largest ocean, and yet their resemblance is uncanny. What can we say based on these facts? At a minimum, we can say that this lineage of crane is very, very old, given the time required for these wild animals to evolve along separate paths to become distinct species and then migrate across a sea and huge landmass. As it turns out, the red-crowned crane is the common ancestor to both species; the whooping crane is a subbranch.

Originally, I was under the mistaken impression that the whooping crane and red-crowned crane both descended from a common ancestor that spread its range from Alaska across the Bering Sea to eastern Siberia and Hokkaido. Then, I presumed, something happened to separate two groups of the common ancestor of this ancient species of crane, sending the groups off on different evolutionary paths. My initial guess was that this separation event was an Ice Age, but in fact the origins of both species lay far deeper into the past than that.

According to a 2019 academic study of the whooping crane, the red-crowned crane is actually a "parent" or common ancestor to the whooping crane and other crane species. Genetic studies show that the species *Grus japonensis* branched off from other crane species somewhere from 7.6 to 9 million years ago. The whooping crane originated from prehistoric red-crowned cranes, branching off through the process of speciation about 4 million years ago. Hence the uncanny similarity between the two species.

Like the original human inhabitants of the New World, the whooping crane is a migrant from Asia. "The historical biogeography of cranes has never been studied, but the ancestral area reconstructions would almost certainly find that the Americana group arose in Asia, with the whooping crane dispersing into North America during the Pliocene, presumably via the Bering Land Bridge that was open intermittently during the Cenozoic," writes Carey Krajewski, a zoologist at Southern Illinois University. Krajewski adds to the evidence the discovery of fossils of whooping cranes dating to the Pliocene.[30] Still, whooping cranes and red-crowned cranes appear nearly identical today; the only visible sign differentiating them is the black plumage on the feathers and necks of the red-crowned crane. Both species have "red crowns," and my Japanese students actually have trouble telling the two apart.

They're both large wetland birds with omnivorous diets, similar coupling behaviors, and similarly slow breeding cycles. Both were nearly driven to extinction by the 1920s due to hunting and habitat loss. Both species are legally protected and subjects of active government-led recovery efforts.

Nevertheless, there are important differences between these two species that can't be ignored and should be highlighted in a review of these two conservation stories. Any comparisons between the species, similar or no, should be approached with caution, given that the differences between these two species could be just as important as their similarities in terms of ascertaining how and why the Japanese red-crowned crane and AWB whooping crane populations recovered at such radically different rates. Four key differences between these two species are discussed below.

First, whooper, the Aransas-Wood Buffalo or AWB whooping crane, is a fair-weather bird—it summers in Canada and winters in Texas. I may retire like this someday. By contrast, cousin *tancho*, the red-crowned crane of Hokkaido, is clearly adapted to colder and more extreme climates. The annual mean temperature at Kushiro Marsh National Park is just 6 degrees Celsius according to Japan's Ministry of the Environment. That's about 43 degrees Fahrenheit. Temperatures at Kushiro Marsh routinely drop below minus 20 degrees Celsius during winter nights, or minus 4 degrees Fahrenheit. The Hokkaido resident red-crowned crane population remains in this same, mostly cold region, prone to temperature extremes year-round and frigid winter conditions. The region receives less annual snowfall than other parts of Hokkaido, but heavy snows can and do occur with relative frequency. This hard winter existence is partly ameliorated by the geothermal activity in Kushiro Marsh, where natural hot springs keep parts of Kushiro River and other smaller tributaries ice-free year-round. However, the red-crowned cranes generally cannot escape bitter cold conditions for several months of the year, and yet they are thriving in terms of population size compared to the AWB whooping crane population.

Second, the AWB whooping crane displays extreme migratory behavior, completing a journey of some 2,500 kilometers (more than 1,500 miles) from their summer breeding grounds in Canada to their wintering grounds in Texas twice a year. Meanwhile, the vast majority of Japanese red-crowned cranes keep to their primary habitat in East Hokkaido in and near Kushiro Marsh National Park. The greater energy expenditures necessary for the AWB whooping crane to make such an epic journey every year could render that population more vulnerable to mortality factors, namely disease or predation or both, meaning it's reasonable to expect that this migration could result in higher mortality compared to red-crowned cranes. However, the existing academic literature has so far determined this is not the case.

The time required for the whooping cranes to complete their migration also means that the whooping cranes spend a significant amount of time outside areas staffed by conservation authorities and thus beyond the protection of wildlife managers. In Japan, MoE has the advantage of being able to intervene in crane protection and conservation as it sees fit at any time of year, since most Japanese red-crown cranes spend the entire year at either the government reserves or protected feeding stations. Even when feeding in farmers' fields, the birds are never too far away. Government conservation workers and researchers in Texas and Hokkaido both rely on many of the same population management techniques, including banding and radio tracking, health checks, and occasionally veterinary care, but MoE officials can conduct these research and care activities year-round, whereas officials at the Aransas National Wildlife Refuge have a window of only about four months to conduct this kind of population health monitoring activity. The Canadian authorities take over when whoopers spend time in Wood Buffalo National Park. When the whooping cranes make their way between these two bases, they are monitored, but conservationists mostly keep their hands off the flock unless a bird happens to be injured.

Third, the AWB whooping crane is notoriously shy. That's why I called the whooper the antisocial crane. I still haven't experienced a truly close encounter with one, and yet here I am writing a book about the species. The best way for visitors to see AWB whooping cranes at the Aransas National Wildlife Refuge is through boat tours departing from a dock in the community of Rockport, but even this doesn't guarantee that one will get close enough to take a clear picture (I should know). The situation is quite the opposite at Hokkaido red-crowned crane winter artificial feeding stations. The grateful crane really does pay back its gratitude in a sense. During winter feeding, Japanese red-crowned cranes have been seen to be extremely habituated to human presence. The cranes are still careful to avoid getting too close to humans, and red-crowned cranes will escape if a human attempts to approach them, but they are much less shy and skittish around people compared to AWB whooping cranes.

During the feeding round that I observed at the Itoh sanctuary, red-crowned cranes stood within 10 meters of a crowd of photographers, some occasionally awing the crowd by flying directly overhead, casting long shadows over the visitors and their vehicles. Japanese red-crowned cranes routinely disrupt automobile traffic in and around Kushiro Marsh while crossing roads. My experience wasn't unusual at all. Red-crowned cranes also are not shy about invading human infrastructure, even when humans are present and working. A penchant for avoiding humans and human infrastructure, including the massive new wind energy arrays being built or already erected in the AWB whooping crane migration corridor,[31] should afford AWB whooping cranes better chances of survival compared to red-crowned cranes willing to dodge traffic and play chicken with farm equipment.

How about their respective habitats? In total area, the Aransas National Wildlife Refuge in southeastern Texas is twice the size of Kushiro Marsh National Park, although the national park is contiguous and the national wildlife refuge is a collection of adjacent parcels. Meanwhile, Wood Buffalo National Park in Canada is approximately the size of Connecticut; it's Canada's largest national park. No one can say that the whooper is hurting for available habitat, at least in my humble opinion. And the AWB whooping crane's habit is expanding still. The adjacent Powderhorn Ranch was taken out of private hands for the purpose of government-driven whooping crane conservation initiatives, but this time the Texas state government wildlife authorities are joining the fray.

Tancho enjoys no such luck. It's theoretically possible for the Ministry of the Environment to expand the size of Kushiro Marsh, but it would have to purchase large tracts of land to do so. MoE is actually the weakest and most poorly funded of Japan's government ministries, so this is unlikely to happen in the near term. Meanwhile, the available habitat in the marsh most suited to red-crowned cranes has contracted over the years due to the excessive sedimentation and Japanese alder encroachment problems. Though they have far less land and available habitat on which to spread their wings, the red-crowned cranes have leapfrogged the whooping cranes of Texas and Canada in population growth.

Finally, red-crowned cranes display more communal behavior. Though they migrate together, AWB whooping crane couples spend their time at the Aransas refuge and in Wood Buffalo National Park apart from other couples. By contrast, Japanese red-crowned cranes gather in the hundreds to take advantage of winter artificial feeding stations or to hunker down for cold winter nights in shallow, hot-spring-fed streams. The red-crowned cranes do space themselves farther apart when breeding in the wetlands of eastern Hokkaido in the warmer spring and summer months but not to the extent of the whooping cranes. Here again one might argue that such circumstances should be beneficial to AWB whooping crane survival, given that crowding pressures food availability during the nine months of the year when red-crowned cranes must forage for themselves. Crowding also leaves red-crowned cranes more susceptible to disease outbreaks, but thus far the population hasn't faced any major disease threats. The manager of the Akan refuge told me that they've been concerned about the possibility of avian influenza spreading from farms to the red-crowned crane population. Thankfully, that hasn't happened yet, at least as of this writing. Disease outbreaks are less probable with the AWB whooping crane population given that species' less-social behavior.

Parasitism doesn't seem to be impacting reproductive rates in these two species, either. The numbers and population growth trends of red-crowned cranes alone tell us that parasites, though certainly present in red-crowned cranes, haven't inhibited *tancho*'s reproductive success. Studies also have delved into the threat that parasites pose to whooping cranes. An investigation led by researchers at Texas A&M University examined whether intestinal tract parasites might be causing fatal diseases to occur in AWB whooping cranes, potentially playing a role in higher rates of mortality. They discovered that nearly one-third of the AWB whooping crane population was infected with the *Eimeria* species of parasite, but that infection rates remained relatively stable, with no sign of elevated transmission rates during the study period. They suspect that these parasites and the diseases they bring could cause higher whooping crane chick mortality, but this is inconclusive, since determining a connection would require investigating extremely remote and difficult-to-reach nesting sites in Wood Buffalo National Park. Besides, the researchers noted that this parasite's prevalence is nearly the same for the red-crowned cranes of Japan. "Previous studies in wild red-crowned cranes (*Gus japonensis*) in Japan found a similar level of *Eimeria* infection in fecal samples (26%) and a higher percentage of infection in samples collected in December compared to January through April," the authors noted.[32] If parasites and the diseases they deliver aren't slowing *tancho* down, it's reasonable to assume that they are relatively inconsequential as far as whooper recovery is concerned.

So what does all this tell us?

Various differences inherent with these two species could be shown to be both detrimental and beneficial to the survival of either. For example, a preference for warmer climates should ensure more year-round and steady wild food availability for AWB whooping cranes, and this shy species behaves in ways that limit exposure to disease, predators, and humans, precautionary behavior that is no doubt beneficial to its survival. The more human-habituated Japanese red-crowned crane takes more chances and consequently could be suffering higher rates of mortality. Yet today the red-crowned crane population is more than triple the size of the AWB whooping crane population, even though the two populations were at comparable numbers of surviving individuals back in 1950.

The AWB whooping crane endures a massive and taxing migration twice each year. This could cause higher rates of whooping crane mortality compared to the more sedentary red-crowned crane of Hokkaido. However, there is little indication in the academic literature that the extreme migration route taken by AWB whooping cranes leads to higher mortality and thus negatively impacts that population's growth rates. Research has demonstrated that AWB whooping cranes actually avoid possible threats that they encounter along their migratory path, including wind towers and other human infrastructure.

The various behavioral differences in these particular separate species, while important, appear to largely cancel each other out, at least for purposes of this particular discussion. They are behavioral differences that can be said to be advantageous or disadvantageous to the population growth for either of the two species.

Endangered species conservation strategy matters, and we can discern patterns or lessons from analyzing these separate approaches to managing these two geographically separated species, which are nevertheless extremely close physiologically, biologically, reproductively, and behaviorally.

These two management regimes have been undertaken by wealthy governments as well-financed initiatives for approximately the same length of time, decades-long initiatives led by highly trained wildlife managers. These case studies are an opportunity to examine whether an emphasis on habitat management versus direct population management via artificial feeding yields faster, more long-lasting endangered species recovery.

Americans of my generation were raised to hold dear to their hearts a core principle of nature conservation and human-wildlife interaction: you do not feed the wildlife. To survive and thrive, wild animals must be forced to learn how to forage on their own, we're taught. Feeding wildlife can also be dangerous, usually for the animals. When a park ranger shoots a black bear, it's often because the bear has learned to associate humans with food and thus poses a threat. They say they do it to protect humans, and I believe them, but it's the bear that ends up dead. To avoid these tragic situations, we are told to observe and admire wildlife but also to keep our distance and to never, ever toss them food.

Here, I return to a central question of this book: why did the red-crowned crane population rebound so much more strongly than the AWB whooping crane population, even with the power of the Endangered Species Act behind the whooping cranes? I've been mulling this question for years, and I keep coming back to the best and most likely explanation: Japan's famous crane winter feeding campaign.

What are the lessons learned from the Endangered Species Act here? Mainly, that the ESA isn't the only game in town. I also would add that wildlife managers armed with the Endangered Species Act possibly haven't been curious enough to study and incorporate lessons from other highly visible, highly successful extinction avoidance and species recovery programs happening in other countries, even when these success stories involve endangered species that are nearly identical to those in North America.

In this chapter, I chided the Japanese authorities for not considering ways to reduce erosion and sedimentation from the farms surrounding Kushiro Marsh, which they could accomplish by making an ally out of an ostensible enemy: Japanese alder. This seems to me an obvious and easy intervention to undertake. Earlier, I chided the press corps for its neglect of what's arguably the most important story of our generation: the sixth mass extinction event now underway warrants at least as much daily media attention as global warming.

Now, I chide American wildlife managers for their failure to learn and incorporate lessons that are plain as daylight and visible for anyone to see—lessons stemming not from the implementation of the Endangered Species Act, but from places and endangered species recovery programs that never benefited from the act. Many of you reading this now may find this characterization unfair, and you have a point, I suppose. After all, wildlife managers in the United States cooperate and coordinate with colleagues in other nations, too. But I find little indication that the American authorities truly and comprehensively have reviewed species recovery successes in other jurisdictions for policy-making purposes. They're binding themselves to the ESA and their own experiences and expertise while possibly overlooking practices or strategies successfully implemented in places where the ESA holds no sway.

I argue here that the ESA's stewards should more actively seek lessons from the broader world, not just North America. Because, to tell the truth, there likely aren't too many places left in the world that haven't been impacted by the Endangered Species Act in some way, I suspect.

You see, the Endangered Species Act wasn't transformative only in the United States. The ESA's shadow looms large over the entire world. It's one of the best examples, in my view, of a local solution making a big impact on a global problem.

CHAPTER SIX

~

The World's Endangered Species Act

Why was it that some 50 years ago the U.S. Congress felt compelled to draft a lengthy law detailing precisely how the U.S. federal government would pursue protections for endangered wild plants and animals? We could ask Joe Biden this question. I'm not kidding—he was sworn in as senator from Delaware in January 1973, becoming the youngest person to serve in the Senate at the time. He was there when the Endangered Species Act was debated and adopted later that year. But we don't need to guess or ask a president for Congress's reasons for drafting and voting the bill into law back then—the lawmakers serving at the time (perhaps many of them still are in office, not only Biden) spelled it out for us in the Endangered Species Act itself. The text of the law begins as follows:

> SEC. 2. (a) FINDINGS.—The Congress finds and declares that—
>
> (1) various species of fish, wildlife, and plants in the United States have been rendered extinct as a consequence of economic growth and development untempered by adequate concern and conservation;
> (2) other species of fish, wildlife, and plants have been so depleted in numbers that they are in danger of or threatened with extinction;
> (3) these species of fish, wildlife, and plants are of aesthetic, ecological, educational, historical, recreational, and scientific value to the Nation and its people.

And because of this:

> (b) PURPOSES.—The purposes of this Act are to provide a means whereby the eco-systems upon which endangered species and threatened species depend may be conserved, to provide a program for the conservation of such endangered species and threatened species, and to take such steps as may be appropriate to achieve the purposes of the treaties and conventions set forth in subsection (a) of this section.
>
> POLICY.—(1) It is further declared to be the policy of Congress that all Federal departments and agencies shall seek to conserve endangered species and threatened species and shall utilize their authorities in furtherance of the purposes of this Act.
>
> (2) It is further declared to be the policy of Congress that Federal agencies shall cooperate with State and local agencies to resolve water resource issues in concert with conservation of endangered species.

In short, Congress collectively "found" that far too many amazing species once native to the territory of the United States had been driven to extinction by human beings and that far too many more were at risk of suffering the same fate back in the 1970s and for the same reason that they were so threatened in the 1870s. Congress also found that species at risk of extinction were of value to the citizens of the nation for the ecological functions they provide and for various other educational, recreational, scientific, and aesthetic reasons—losing these species would make the United States less beautiful and less intellectually and culturally rich, in other words.

Thus, the "purpose" of the U.S. Endangered Species Act (ESA) is to put a stop to this killing and death and even reverse it, if possible, to preserve these species for the benefit and enjoyment of future generations. The ESA was also enacted as a vehicle to spell out exactly how the U.S. government planned to comply with various bilateral and multilateral treaties relating to endangered species protections that the federal government had already signed and ratified, as Congress noted in the beginning paragraphs of the law. By drafting this landmark piece of legislation, Congress declared it the "policy" of the federal government that every federal agency and department cooperate toward the protection and conservation of endangered species, to both improve the domestic condition of the United States and to meet America's obligations with respect to other nations. The law later clarifies that ESA enforcement power rests with the Department of the Interior.

A government finding that species are being driven to extinction by humans. A government deciding to do something about it. And that same government determining that species' conservation will be an explicit policy objective to be pursued for domestic reasons and because that government is already obliged to comply with endangered-species-related international treaties. This is the overall starting tone of the Endangered Species Act, its raison d'être. Other countries' endangered species laws start out on the same foot. They also mimic or emulate the ESA in many telltale and extremely interesting ways, a clear sign of the tremendous impact that the '73 act has had on the world.

~

As I stated earlier in this book, today endangered species conservation and recovery policy is at a historic turning point, in the United States and for the world as a whole.

December 2023 marks the 50th anniversary of the passage of the 1973 Endangered Species Act of the United States. This landmark act by the U.S. Congress, and a main focus of this entire book, was made possible through strong bipartisan political cooperation on environment protection, capping off a string of strong progressive legislative victories that would set the tone for U.S. environmental policy at local, state, and federal government levels for decades. But the ESA didn't just transform biodiversity protections in the United States. Its influence reverberates well outside America's shores.

Beyond its domestic significance, the 1973 Endangered Species Act had a clear and recognizable impact on global environmentalism by influencing and inspiring endangered species legislation in a host of other nations, as well as the 1992 United Nations Convention on Biological Diversity. How did it accomplish this? In part, it achieved historical significance because at the time of its passage, the U.S. Endangered Species Act was far and away the most comprehensive piece of legislation ever drafted with the aim of halting and reversing rising rates of species extinctions and biodiversity devastation. Other nations later emulated the ESA in their past and presently enforced endangered species laws and regulations, even as late as 2019. The result was the eventual emergence of a set of global standard practices for species and habitat protections that endure to this day, practices with obvious American roots.

What follows is a deeper dive in the legacy and impact of the ESA outside America's borders. I adhere to a certain academic writing style in the following pages, so please bear with me as you work your way through this chapter. I still try to keep things interesting here, though.

Below, I explore in more detail the degree to which the past and present endangered species legislation of other English-speaking jurisdictions borrowed from or mimicked the U.S. ESA in purpose, style, form, and function. I also undertook some further textual analysis to reveal as much as I could about how much Japan's own endangered species protection statute is influenced by the U.S. ESA. This exercise involved combing through a whole lot of incredibly boring and terribly written legalese for hours at a time (unfortunately but inevitably, legislation is drafted by lawyers, just as treaties are drafted by diplomats, to the detriment of readers everywhere). Still, by bearing this burden, I was able to discover or at least reveal the profound influence that America's 1973 Endangered Species Act has had on endangered species protection practices in other countries for decades.

Soon after its initial passage, the ESA became a sort of template for foreign national and provincial governments to follow or emulate when mulling and eventually drafting or redrafting their own endangered species management protection programs. The ESA's passage and its emulation by other governments is, in my opinion, indicative of the degree to which the United States has influenced and continues to influence global environmentalism. We often hear complaints, particularly from progressive corners, that the United States too often avoids assuming a leadership role in global environmental concerns. This is an especially common complaint when the issue at hand is climate change—note the U.S. withdrawal from the Kyoto Protocol and President Trump's quitting of the Paris Agreement, later reversed by President Biden. But when it comes to endangered species management and conservation, in many important ways, America forged the mold.

But to better understand how and why, we should first take a deeper dive into international relations theory and some foundational aspects of how the world pursues cooperative environmental management and protection at the United Nations and on other multilateral stages. I bet you weren't expecting this turn. Neither was I when I began writing this chapter, but there's value in pausing to consider how and why nations get together to cooperate on the environment (or fail to, as is frequently the case).

Since the 1970s, international environmental diplomacy has developed a discernible pattern that helps to explain how the current global natural resources management regime or mechanisms emerged and evolved. In other words, global environmental protections and natural resources management at the United Nations and beyond have developed precise characteristics that can be clearly identified and parsed. Understanding these characteristics and identifying how and why international environmental agreements are formulated the way they are (due to the realities of modern international affairs) helps us better understand why the Endangered Species Act has had such a tremendous impact on the globe.

There are many flavors and layers to international relations theory, or how international relations theory is defined and taught. I won't get into all of them here. For purposes of my research and teaching, I focus on three general theories regarding how nations pursue relations with one another: realism, institutionalism, and constructivism.

International relations (IR) scholarship begins with a simple and obvious premise: there is no one world government, meaning no single entity or authority ruling over the entire planet. In the animated television series *Futurama*, Earth is ruled by a single Earth government, and all humans are citizens of Earth—the governments of the United States, China, Canada, Uganda, Peru, New Zealand, and on and on are all things of the past. In the *Star Trek* television series, Earth belongs to an even larger governmental organization called the United Federation of Planets. In our reality, Earth's surface has been broken into pieces by nearly 200 governments, each claiming their own little or large slice. There is no "Leviathan," as the English philosopher Thomas Hobbes put it—no single, overarching, dominate authority that can enforce peace and cooperation among the disparate nations of our planet. The United Nations is not a single world government; it exists solely because separate sovereign and independent nations agree to gather in large auditoriums or around tables in New York or Geneva and try to hash things out as best they can. Thus, in a nutshell, the study of international relations is a study of how nations compete or cooperate in an environment where there is no overarching authority or power head. It's a study of how states go about their business in a (more or less) state of anarchy.

Realists, or proponents of realism theory, assert that this condition of anarchy forces states into a condition of mostly competition. In other words, realists insist that states are constantly competing with each other and that this is the driving reality underpinning global affairs. Power is paramount, and states navigate turbulent waters as best they can by pursuing alliances and occasional wars in a never-ending attempt to maximize their gains and minimize their losses. Thus, when Russia invaded Ukraine in early 2022, realists might assert that this occurred because Russia sees itself in competition with the NATO alliance. Fearing a relative loss or diminishment to its power vis-à-vis NATO, Russia launched an unprovoked attack on a neighbor that didn't threaten it in a bid to tip the scales of the regional power balance a little bit more in its favor. And Russia took this risk because its leadership calculated that it had more to gain than lose from pursuing an attack on Ukraine. That's sort of how realists see the world we live in.

Institutionalists, or adherents to institutionalism theory, are more optimistic or rosy in their assessment of the world than the realists. Institutionalism argues that states understand perfectly well that they must not only compete, but also cooperate. There is no Leviathan, and no single nation is powerful enough to become the dominant single legal authority on Earth, so in a bid to avoid excessive and potentially disastrous competition (such as a war in Europe in the 21st century), states endeavor to form international institutions through which they can pursue peace and cooperation instead. As proof of this reality, intuitionalists can point to the United Nations itself. Even though the League of Nations fell apart with the start of World War II, nations dusted themselves off and tried again with the creation of the United Nations in 1945. That global forum for cooperation and peacekeeping exists to this day.

Constructivism is somewhat more compelling, in my opinion. Constructivism theory places special emphasis on the power of ideas and norms. To put it another way, constructivism makes note of the fact that people think differently today than they did 1,000 years ago, 100 years ago, 50 years ago, or even just 10 years ago. What was once considered normal and natural in the past is no longer deemed acceptable to people living in the present, and people ultimately run international affairs. A classic example here is slavery. For thousands of years, putting other humans in bondage and treating them as chattel were deemed acceptable and even natural practices. Thankfully, this is no longer the case. That's an extreme example, and there are many more subtle ones—women's suffrage, for example. The main point of constructivism is that the thinking of entire populations changes over time. In more recent times, constructivists say that nations have endeavored to show that they are good, responsible actors, or actors willing to assume a constructive or more productive role in global affairs. This isn't true for every nation, of course (think North Korea). But it does hold true for a substantial number of them (think Canada).*

Starting with these three principal theories of international relations—realism, institutionalism, and constructivism—we can begin to understand why the world developed the global environmental protection or natural resources management architecture that it has.

Namely, through the United Nations and other fora, the international community has adopted a set of key international environmental treaties. These treaties perform one or more of three basic functions: they assign

---

* The fundamentals of international relations theory are laid out nicely by Kate O'Neill of the University of California at Berkeley in her book *The Environment and International Relations* (Cambridge: Cambridge University Press, 2017).

Red-crowned cranes in Hokkaido, Japan. *Atsuko Ellie Teramoto, Ellie Teramoto Photography.*

Red-crowned cranes dancing in East Hokkaido. *Atsuko Ellie Teramoto, Ellie Teramoto Photography.*

Whooping cranes in Kentucky. ***U.S. Fish & Wildlife Service.***

Whooping cranes at the Aransas National Wildlife Refuge in Texas. ***U.S. Fish & Wildlife Service.***

Attwater prairie chickens in a protective enclosure at Attwater Prairie Chicken National Wildlife Refuge, Texas. *Atsuko Ellie Teramoto, Ellie Teramoto Photography.*

Attwater prairie chickens in a protective enclosure at Attwater Prairie Chicken National Wildlife Refuge, Texas. *Atsuko Ellie Teramoto, Ellie Teramoto Photography.*

A wolf pacing through the trees at the Lakota Wolf Preserve, New Jersey. *Atsuko Ellie Teramoto, Ellie Teramoto Photography.*

Sandhill cranes at Bosque del Apache National Wildlife Refuge, New Mexico. *Atsuko Ellie Teramoto, Ellie Teramoto Photography.*

European white stork. *Photo by Jan Stefka, Creative Commons.*

California ridgeway rail. *U.S. Fish & Wildlife Service.*

Mauritius parakeet. *Photo by Peter R. Steward, Creative Commons.*

Houston toad. ***U.S. Fish & Wildlife Service.***

Tasmanian devil. ***Photo by Mathias Appel, Creative Commons.***

Steller's sea eagle in East Hokkaido, Japan. ***Atsuko Ellie Teramoto, Ellie Teramoto Photography.***

Vaquita porpoise in the Gulf of California (Sea of Cortez), Mexico. ***Photo by Thomas A. Jefferson, Creative Commons.***

Bald eagle perched on a tree near Houston, Texas. ***Atsuko Ellie Teramoto, Ellie Teramoto Photography.***

California condor in Grand Canyon National Park. National Park Service. ***Creative Commons.***

rights to resources or rights to access areas of potential resource discovery, they facilitate information-sharing regimes to enhance cross-border environmental protections or rules enforcement, or they encourage (but never force) sovereign states to protect or sustainably manage resources found on their territories.

The United Nations Convention on the Law of the Sea (UNCLOS) is a primary example of how states have opted to assign rights to resources or resource extraction areas to one another. Cobbled together from earlier maritime-related agreements and practices, including a few new ones, UNCLOS was a way for nations to overcome the realist problem of potentially dangerous competition for resources while balancing the necessity of freedom of navigation of the seas. Negotiators did so by drafting this massive treaty in a way that gives nation-states exclusive rights to economic exploitation in certain zones of the oceans, while maintaining the rights of other states to navigate their vessels through these same zones. The largest area of assigned rights codified in UNCLOS is the exclusive economic zone or EEZ, a zone extending 200 nautical miles from a nation's shores. In the EEZ, that sovereign nation retains exclusive rights to maintain or manage the resources found within it however it sees fit. Some states might ravage their EEZs, nearly emptying them of fish. But others choose more sustainable management practices to avoid future economic pain. However, the EEZ is *not* a nation's territorial sea zone, and this point can't be emphasized enough (though the government of China has attempted to claim as much in the past, ignoring its own obligations as an UNCLOS party state). You and I and everyone else are free to traverse any EEZ in any part of the globe in any vessel and at any time. We're just not allowed to exploit any of the resources in these zones or generally make money from them.

EEZs are not only about fishing. The real prize is often offshore oil and natural gas, as well as occasional mineral wealth. Nations can appeal for further exclusive rights to exploit the ocean bottom through the Commission on the Limits of the Continental Shelf, an advisory body established by UNCLOS. Moving further out of the EEZ, an extended continental shelf puts one in the high seas, a zone owned or exclusively managed by no single state. But UNCLOS assigns rights to resources found in this region, as well.

The treaty established the International Seabed Authority (ISA), long an obscure UN agency that meets in Kingston, Jamaica, every once in a while, but one that you will hear about much more about in the near future. ISA has identified those parts of the high seas ocean floor that show the greatest potential for mineral wealth extraction. It's since divided the international seafloor into sections, licensing states to manage or exploit these sections at some point in the future—assigning rights, in other words.

Other international treaties set up complex information-sharing regimes in a bid to help governments head off environmental threats, including threats to endangered species. Arguably the most powerful treaty that protects the environment through information sharing is the Convention on International Trade in Endangered Species of Wild Fauna and Flora, better known by its acronym, CITES. Through CITES, governments apply to list species they feel are threatened with extinction. This list is shared with other governments, and all parties have agreed to use this information to strictly monitor and control cross-border trade in live plants and animals and their parts, especially if a species is deemed threatened or endangered in another nation. Rare for a multilateral environmental treaty, CITES has teeth—nations are loath to fall out of compliance with CITES, both because it would be embarrassing and because the agreement authorizes serious trade consequences.

The Basel Convention on the Control of Transboundary Movements of Hazardous Wastes and Their Disposal is another classic example of the information-sharing function of some international environmental treaties. The Basel Convention commits state parties to pass laws requiring companies or individuals operating in their countries to inform other countries about incoming shipments of hazardous waste prior to those shipments leaving a home port. This essentially means that if a company in country A wants to ship some petroleum-refining waste to country B, country A's laws require that company to first inform and receive consent from the government of country B before that waste can leave port. Country B can agree to accept the waste—maybe there's money to be had in disposing of it—but it can also reject the waste shipment, thereby potentially preventing a major environmental problem from happening. Stiff fines and possibly even jail time await the officers of that company should they choose to ignore the Basel Convention and get caught.

Other global agreements function by encouraging nations to protect their natural resources, or at least to not overexploit various natural resources existing within their territories. A great example is the World Heritage system run out of the United Nations Educational, Scientific, and Cultural Organization (UNESCO). Through the World Heritage Treaty, nations initially sought to develop a mechanism for protecting antiquities and historic sites but later broadened the concept of "world heritage" to include nature as well. Today, the moniker "UNESCO World Heritage Site" is a gold standard for protected areas. The treaty invites nations to nominate regions of stunning natural beauty and significance to earn the title of World Heritage Site. It's the brand that keeps giving—studies show a significant bump in tourism visitations and tourism revenues to localities and governments after a national park or protected area achieves UNESCO World Heritage status. The fancy title also attracts scientists, research funds, and independent conservation initiatives, augmenting a state's own capabilities. But the fancy name comes with strings attached. To keep a protected area's World Heritage status, governments must adhere to a set of conservation rules and best practices. Failure to do so means that nation could suffer the humiliation of having a site's World Heritage status revoked. This has happened before. Thus, UNESCO encourages environmental protection through this creative and highly effective marketing mechanism.

That's basically how global environmental management works: states have cobbled together a series of agreements that (a) assign rights to resources or rights to access areas of potential resource abundance, (b) facilitate information-sharing regimes to augment government capacities, or (c) encourage states to better protect the environment within their own sovereign territory. This will all be on the final exam.

∼

Now, back to bare-bones, simplified international relations theory.

UNCLOS shows bits of realism and institutionalism at play. Initially, states would unilaterally declare 200-nautical-mile exclusive economic zones around their shorelines (the United States did so for quite a number of years). Realists would note how powerful states could make and enforce these unilateral sea grabs. But ultimately, institutionalism won the day, and states sat down and cooperated to draft a treaty and form institutions like the International Seabed Authority and the Commission on the Limits of the Continental Shelf to better organize claims and avoid excessively damaging competition.

Institutionalism perhaps came to the fore again when nations negotiated CITES and the Basel Convention, the information-sharing regimes. It's not too difficult to imagine how governments stand to benefit tremendously by sharing information with each other, making all participating governments smarter and more aware of what's happening than would otherwise be the case. Here, cooperation makes way more sense than competition.

At UNESCO, I see both the institutionalists and constructivists flexing their muscles. Hosting one or several Natural World Heritage Sites obviously has great appeal, not to mention is an excellent way to advertise your country to would-be foreign travelers. It's the sovereign governments that take the initiative to set aside areas for conservation and then appeal for UNESCO recognition. They bother to do all this, constructivists argue, because nations run by people existing in our modern reality want to be seen as good stewards of nature and responsible actors in global conservation efforts; otherwise, they would simply ignore UNESCO and rip up verdant mountain ranges and valleys and coastal areas to their heart's content.

So how does the U.S. Endangered Species Act fit into this framework?

I already hear you scoffing. You must be thinking, "But hold up, the ESA is not an international environmental treaty!" And you're right. The ESA is a piece of domestic legislation. It was drafted and passed by and for the U.S. government alone. It was never intended to be an international legal document, inspired or otherwise.

And yet, in a way, the Endangered Species Act actually has gone international, and it did so a long time ago. Perhaps it was always destined for international greatness or significance.

The U.S. Endangered Species Act isn't international law per se, but in a way, it kind of is. Allow me to explain.

In January 1973, an initial draft version of the U.S. Endangered Species Act was brought to the floor of the U.S. Congress. According to the historical archives of the U.S. House of Representatives, lawmakers would spend the next 12 months debating back and forth and revising the act before agreeing on its basic form and function. In December 1973, a finalized version of the Endangered Species Act passed the Senate with near-unanimous approval. On December 20, 1973, the U.S. House of Representatives approved the act with overwhelming bipartisan support, voting 355 to 4 in favor of its passage and sending the bill to President Richard Nixon for his signature. President Nixon signed the U.S. Endangered Species Act into law on December 28, 1973, per congressional records. The law supplanted the Endangered Species Conservation Act of 1969, which itself was an expanded version of the earlier Endangered Species Preservation Act of 1966.

The '66 act first established the Department of the Interior's mission and authority to prevent species' extinctions. Congress authorized funds for endangered species protections with that initial act's passage, and the Department of the Interior issued the first official U.S. government list of endangered species in 1967. That first list of federally recognized endangered species included only a few dozen iconic species at the time. One of them was the whooping crane.

The 1969 act directed the Secretary of the Interior to greatly expand the list of endangered species and to then ban the import and export of species on that list, foreshadowing the soon-to-be CITES (which was adopted by a small number of UN member states some months before the 1973 ESA was finalized). The whooping crane was among the first species listed for protection under the 1969 act, essentially becoming grandfathered into the newer legislation, and the crane's status as an endangered species to be preserved was carried over to the new list of protected endangered species under the newest and currently in force act, the 1973 ESA.

I'm being repetitive here, but it's worth emphasizing the point, I think. At the time of its adoption by Congress and signature by President Nixon, the U.S. ESA was the most comprehensive piece of legislation addressing protections for wildlife and their habitats ever enacted by any government. During the next several years, other governments at the state, provincial, and national levels would more or less follow America's lead. These laws were either designed for the same purposes as the ESA or as instruments through which governments planned to implement CITES, a powerful multilateral environmental treaty that preceded the ESA, which the ESA itself references. There are numerous examples of this trend.

The Wildlife Act of Ireland was enacted in 1976 (and later amended in 2000). Norfolk Island, then a self-governing territory of Australia, passed its own Endangered Species Act in 1980. The United Kingdom enacted the Wildlife and Countryside Act in 1981. The Canadian province of Manitoba passed its Endangered Species and Ecosystems Act in 1990. Australian central government lawmakers enacted that nation's comprehensive Endangered Species Protection Act in 1992. Lawmakers in the Philippines adopted Republic Act No. 9147 in 2001, that nation's law governing endangered species protections. Canada's central government implemented the Species at Risk Act in 2002, expanding legal protections for whooping cranes in North America. And perhaps most recently, Uganda's legislature enacted the Uganda Wildlife Act back in 2019. This is by no means a comprehensive list of all national and subnational endangered species laws that followed the 1973 ESA and that were largely modeled after it. Japan's national government enacted the Act on Conservation of Endangered Species of Wild Fauna and Flora (ACES) in 1992, the same year member states of the United Nations adopted the Convention on Biological Diversity. Japan's ACES law entered into force in 1993, 20 years after the Endangered Species Act was passed.

As mentioned repeatedly, enforcement of the ESA is handled through the U.S. Department of the Interior, further delegated to the U.S. Fish & Wildlife Service and the National Marine Fisheries Service. The ESA directs the Secretary of the Interior to carry out assessments of native plants and animals and to determine their status in the wild. The secretary is directed to determine, based on these assessments, whether a species is threatened with possible extinction or at risk of imminent extinction, and then list such categorized species as "threatened" or "endangered" in accordance with ESA procedures as detailed in the law. The ESA then directs the secretary to identify habitat critical to the survival of a species and then to draft a plan for the protection of such critical habitat as deemed necessary to ensure a species is saved from extinction. The process must be transparent and made available for the public to review and comment on. The general public is also invited to propose species for listing and protection.

As I mentioned earlier, the U.S. Endangered Species Act is remarkably international in scope. The ESA specifies that the Department of the Interior must cooperate with foreign governments in pursuing endangered species conservation. The ESA explicitly mentions Japan, Canada, and Mexico as some of the nations with which the United States must cooperate on biodiversity protections per existing treaties. The law directs the Secretary of the Interior to inform foreign governments of listing decisions regarding species also existing on those foreign territories. Foreign governments are even invited to comment on any listing decisions proposed by the U.S. Department of the Interior. The law also says that the ESA is the vehicle through which the U.S. government will enforce provisions of the CITES multilateral agreement. Many other nations' keystone species protection laws do the same.

Globally, endangered species conservation laws are more or less variations on a theme: the United States of America's ESA.

With slight variations, the general species management approach as outlined above and incorporated into the ESA is repeated in form and function in other nations' endangered species protection laws, including at the state, provincial, and national levels, with slight variations and different wording, of course. Indeed, there are important distinctions that can be highlighted between other governments' species protection acts and the U.S. ESA. However, and more broadly speaking, a review of some interesting and specific examples of even a handful of national and provincial legislation that followed the 1973 ESA helps to demonstrate just how much the U.S. government influenced the way the world pursues measures to slow or halt the alarming rate of global biodiversity decline, an ecological disaster that is still very much ongoing.

What follows demonstrates how significantly America's federal government shaped the very foundation of worldwide endangered species management and conservation through its drafting and passing of the 1973 ESA. That influence has been felt and recognized for five decades, and it's likely to endure for decades to come, as governments and the global community continue to struggle to halt the collapse in global biodiversity that scientists and conservationists are documenting daily. To better understand and demonstrate how, I set out in mid-2022 to explore the degree to which other nations have largely emulated or mimicked the letter and spirit of the U.S. Endangered Species Act, to see how the landmark bill essentially forged part of the world's approach toward endangered species management and biodiversity protections since its enactment. In other words, I wanted to see whether or not the essential underpinnings of the ESA and how it works became part of the general scaffolding of global endangered species management.

Indeed, that's what I discovered.

I began my review with a look at how a small portion of the existing academic literature interprets the legacy of the ESA. From there, I took a closer look at the endangered species legislation as passed and enacted by governments in Australia, Canada, Ireland, the Philippines, and Uganda to determine just how similar they are to America's ESA. I downloaded the complete texts of these governments' legislation from the respective official government websites. Most of the time, only PDF file versions of the legislation were made available to download, so I converted all these PDF files into Word format using Adobe's online PDF-to-Word conversion tool to allow for a more thorough textual analysis and cross comparisons.

I also examined, in less fine-tooth-comb detail, the endangered species trade laws as enacted by the United Kingdom, New Zealand, Nigeria, and Singapore, and of course the CITES treaty and the United Nations Convention on Biological Diversity. These were all reviewed separately and in addition to those other examples of national legislation to provide me a more comprehensive picture of the world's approach to biological diversity protection and its roots in American legislation. Initially, I limited this international legislation textual cross analysis to English-speaking jurisdictions, but I later decided to also look closer at Japan's ACES law as well, which I reviewed both in its original Japanese and in a translated English version I found (courtesy of the online service Japanese Law Translation). This let me assess the degree to which Japan's own landmark species protection law may have been influenced by earlier protection models pioneered by the U.S. Endangered Species Act. There are important differences, of course, but the similarities between the two acts are substantive as well.

~

Science has demonstrated how biodiversity itself, or the mere existence of biodiversity, is of direct benefit to humans.

Recent research reveals that bird species' richness strongly correlates to people's sense of well-being. For instance, in areas of Europe believed to be the most biodiverse in terms of bird species, humans living in these same areas reported higher average rates of contentedness and life satisfaction compared to areas of comparatively poor bird species richness.[1] This correlation between bird biodiversity and life satisfaction in Europe is so strong that the additional quantifiable measure of happiness reported by people living in the mere presence of greater bird biodiversity increased in a similar way, as if those same individuals enjoyed increases to their incomes. People like money, of course, but they also like to be surrounded by nature, and one sure sign of nature's presence is birdsong. Even the coldest of concrete jungles can be made more bearable by the singing and chirping of seen and unseen birds.

So humans place a premium on nature and natural places, although we are quite adept at destroying nature. Understanding nature's value, governments everywhere endeavor to some degree to prevent species' extinctions because of the widely held belief (and scientifically confirmed fact) that species richness and greater biodiversity are of benefit to humankind in a variety of ways. The U.S. Congress states this explicitly in the introduction to the ESA and its contents, titled "Findings, Purposes, and Policy." As noted above, in section 2(a)(3), Congress states that species are to be protected from extinction because they "are of aesthetic, ecological, educational, historical, recreational, and scientific value to the Nation and its people." This section avoids mention of the commercial value of species, but it is stated explicitly later in the document that threatened and endangered species are to be both protected from and for commercial activity—economic activity that doesn't directly involve a species should be pursued in a way as to not threaten the existence of a species, the ESA says, and the direct commercial exploitation of a species must be conducted in a manner to ensure sustainable use and the ongoing, long-term existence of said species. The reference to "recreation" also suggests Congress understood back then, as well as today, the commercial potential of biodiversity, in that people will travel and spend quite a bit of money to see and experience species and biodiversity. When I was in Kenya, the government at the time charged foreign tourists $60 per day to visit Amboseli National Park. The locals, of course, paid far less. Sixty dollars per day is a steep price, but Amboseli's managers were never short of customers or pending reservations.

In a way, the ESA introduced to the world the concept of legislating the value of wild species to humans in terms of perceived cultural and aesthetic benefits, for scientific and educational purposes, and for commercial gain. In other words, the ESA legislated the concept of biodiversity as another natural resource for humans to manage, much the same as with mineral wealth, hydrocarbons, or timber. As demonstrated below, other national and international biodiversity laws continued this practice of commodification of wild plant and animal species for management and trade—humans became determined to halt extinctions and take some measures to protect wild species not for the species' sake, but because we convinced ourselves that wild plants and animals are resources to be managed and exploited responsibly by humans. Many environmentalists are appalled by the fact that governments don't protect species for their own sake, but only out of some semblance of commercial and recreational benefit to humans, but that's just the way it is.

How governments pursue endangered species management varies greatly due to variances in national histories, government structures, governing philosophies, and internal politics. Even countries with very similar histories, cultures, shared species, and shared geography can approach the same goals of species conservation in very different ways. For instance, a 2013 review of the differences between the U.S. ESA and Canada's Species at Risk Act (SARA) found the two laws result in the Canadian and American authorities often taking far different approaches in pursuit of the same goals. The researchers noted that the SARA law dictates that all species' status assessments must be undertaken by a single scientific body charged by the national government to perform this work, whereas in the United States the ESA authorizes species assessments by different parties, with Congress declining to use the ESA to forge a single national scientific body that exists only to assess species' status in the wild.[2] At the same time, the ESA mandates strict deadlines for listing decisions and prohibits listing decisions that are influenced by social and economic considerations, meaning the ESA doesn't allow authorities to forego a scientifically sound listing decision simply because it may prove economically detrimental to an individual, group, or company in some way. But they do this anyway, don't they, as we saw with the dunes sagebrush lizard? And the Fish & Wildlife Service (FWS) is accused of doing it again in the case of the lesser prairie chicken, a species laid low by habitat loss and the subject of a recent lawsuit launched in August 2022 by the Center for Biological Diversity—conservation groups have been pressing for an ESA listing for lesser prairie chickens for more than 25 years.[3] One lesson here is that the agencies charged with enforcing the ESA don't follow the law to the letter themselves. On the other hand, Canada's SARA and other national endangered species laws were drafted in ways that make more explicit governments' concerns about the ultimate costs of implementing listing decisions on private interests and communities.

Actions in pursuit of endangered species protections in the United States also tend to be far more litigious than in other jurisdictions, and the ESA itself falls victim to this additional use (or waste) of court time and resources. It's been this way since the law was passed. One famous example is the snail darter case. The snail darter is a tiny fish whose endangered status saw the courts putting the kibosh on a dam project spearheaded by the Tennessee Valley Authority in the 1970s. Only an act of Congress was able to reverse the ruling, but luckily the snail darters survived. The Congressional Research Service (CRS) recently highlighted the sometimes excessively litigious nature of ESA listing decisions and enforcement—and for endangered species management in the United States in general—with a thorough case study of the long, fraught listing history of the gray wolf.

The gray wolf was initially listed as endangered in 1967. More recently, that species has rebounded in certain areas, and the U.S. Fish and Wildlife Service has attempted numerous times to either change its listing status or to delist entirely certain regional populations of gray wolves during the past 20 or so years. As CRS details in its report, lawsuits filed by interested parties have thwarted every single attempt by FWS to delist parts of the U.S. gray wolf population to date, as various environmental groups deftly take advantage of the vague language or wording in ESA sections to block any and all FWS delisting attempts.[4] Though lawsuits over endangered species listing decisions of course happen in many other countries at the national, state, and provincial levels, cases of endangered species listing decisions or other management decisions becoming deadlocked for two decades or longer by contentious and extremely expensive litigation seems to be more of an American phenomenon and one key way in which endangered species management in the United States starkly differs from other nations' approaches.

Still, the similarities with other national endangered species management laws and regimes and the ESA far outweigh any apparent and obvious differences. The earlier report I referenced comparing the ESA with Canada's SARA law (written by Waples and colleagues) argues that endangered species management and protections could be enhanced in important ways in both Canada and the United States if the ESA and SARA were actually redrafted (amended, really) or reinterpreted in certain ways to better mimic or reinforce one another, given that both laws are often directed at the same animal species moving back and forth across the U.S.–Canada border (the best example of this is, of course, the AWB whooping crane, which crosses the border twice annually). In reality, Canada's SARA national endangered species legislation already heavily echoes the ESA in real and consequential ways.

As mentioned earlier, as drafted, the U.S. ESA is both domestic and international in scope. The preamble "Findings, Purposes, and Policy" section references the United States' obligation to international cooperation under migratory bird treaties signed with Canada, Mexico, and Japan, as well as other multilateral wildlife protection conventions—in particular, conventions related to fisheries conservation (Sec. 2a Findings, 4, A–G). The ESA also references the UN CITES convention and U.S. obligations under that treaty. That these mentions appear early in the text of the law make it clear that international environmentalism and rising concern about threats to global biodiversity influenced the drafting of the ESA before its adoption in late 1973—and indeed, the United States itself, as shown by the 1967 and 1969 endangered species laws that the ESA eventually supplanted. Moving beyond 1973, however, the ESA has had a tremendous impact on how other governments frame and pursue endangered species and habitat protection.

The 1973 U.S. Congress specifically charged the U.S. Department of the Interior with the responsibility for enforcing the ESA. The law stipulates that the Secretary of the Interior must make determinations as to "whether any species is an endangered species or a threatened species," with "endangered" defined as facing the imminent prospect of extinction and "threatened" defined as a species declining to such an extent that it may become at risk of extinction in the near future (Sec. 4a General, 1). If such a determination is made, then the secretary is ordered to publicly list that species as threatened or endangered and to subsequently and in a timely manner notify the public of any changes to a listing status that may occur, including adding or removing a species from the list of threatened or endangered species (Sec. 4a General, 2, A).

Thus, a fundamental pillar of government-led endangered species protection worldwide was enshrined first into the U.S. ESA: a designated government entity must make an official determination of a species' threatened or endangered status, then an official listing decision must be formally and publicly announced in a clearly stipulated, procedural way. Any specific changes to an endangered species' listing status (additions, alternations, or removals) must also be made public and in a timely fashion.

Next, the ESA directs the Secretary of the Interior to identify habitats deemed critical to the survival of a threatened or endangered species in the wild (Sec. 4a General, 3, A, i). The ESA excludes from the Interior's jurisdiction properties held by the Department of Defense (Sec. 4a General, 3, B, i)—the military assumes all responsibility for protecting wildlife that exists on its reserves. The step of taking an official identification or determination by the Interior of an endangered species' critical habitat is another unique precedent in species conservation policy enshrined first in the ESA, inspiring other governments to draft and pass laws that spell out procedural designation and management plans for protected areas of land or marine habitat deemed critical to the survival of one or more species. The actual policies pursued can be wide ranging, from basic legal restrictions on activities conducted in these protected areas to an outright government appropriation or purchase of tracts of land or waterways, removing them from commercial considerations entirely. The ESA allows for all of this, and other governments copied this basic approach.

The ESA also greatly empowers the public to drive endangered species management in the United States. This is facilitated by language allowing and even encouraging outside parties or groups to submit petitions to the Department of the Interior requesting that species be listed or delisted as threatened or endangered.

Upon receipt of a petition for a listing decision (addition, revision, or removal), the ESA sets a deadline of 90 days for the Secretary of the Interior to determine whether the third-party petition is warranted, thus prompting a scientific review of a species' status in the wild by the Department of the Interior (Sec. 4b Basis for Determinations, 3, A). The law then gives the Interior another 12 months to notify the public as to how it intends to proceed with the petition for a listing decision (Sec. 4b Basis for Determinations, 3, B). Another common feature of the ESA is the strict timelines it established for listing decisions to be made and announced and the strict requirement that listing decisions be made public (in the *Federal Register* and in local newspapers where a species is known to be prevalent) and that the public be invited to review and even participate in the listing process during a public review and comment period. The ESA attempts to enforce transparency, accountability, and public participation in listing considerations and determinations. And as noted earlier, listing actions or petitions also fall under judicial review, and court challenges are common.

All these core principles and concepts underpinning government-led species protections—procedural, participatory, and timely endangered species listing decisions and policies, always subject to judicial oversight—are enshrined in other nations' endangered species management laws and regulations, as I demonstrate in greater detail below, beginning with the ESA's Canadian girlfriend, SARA.

Canada, 2002 Species at Risk Act

Though Waples and his colleagues argue that Canadian and American endangered species laws are too different and should be better harmonized, a more careful reading of Canada's national 2002 Species at Risk Act (SARA) reveals that lawmakers in Ottawa were inspired by both the letter and spirit of the U.S. ESA in drafting their own foundational endangered species management act. In my reading, the two acts have a lot more in common than Waples and colleagues allude to in their analysis.

Whereas the ESA designates the Department of the Interior as the responsible enforcement authority, SARA directs authorities to form a Canadian Endangered Species Conservation Council consisting of ministers of three federal agencies (in this case, Environment, Fisheries and Oceans, and Parks Canada) and ministers from concerned provinces or territories (Composition 7[1]). Enforcement obligations can be delegated to a specific ministry after consultation with other parties to the council, with a delegation determination made public within 45 days from commencement of a species status review (Responsibility of Minister, 8). The council must also act in consultation with representatives of First Nations groups where endangered species management decisions may impact them (National Aboriginal Council on Species at Risk, 8).

SARA establishes a clearly defined listing process to be undertaken by the Committee on the Status of Endangered Wildlife in Canada, or COSEWIC (Establishment, 14), another body established by the act itself. In this section, SARA goes beyond ESA in listing specificity, authorizing COSEWIC to make a determination on whether a species is "extinct, extirpated, endangered, threatened, or of special concern." These are the legally specified subcategories of the more general "species at risk" classification under Canada's comprehensive endangered species law. COSEWIC is also ordered to note publicly whether there is no cause for concern over a species' status or in cases in which scientific data are lacking to make any specific determination (Functions, 15).

A listing and then a public declaration or public notification—steps familiar to anyone with even a passing familiarity with the U.S. Endangered Species Act. Under SARA, COSEWIC is directed to establish subcommittees to undertake reviews of the status of individual species (Subcommittees, 18[1]) and as with the ESA, SARA specifies a timeline for action: the law gives COSEWIC one year upon receipt of a subcommittee report to make an assessment and listing decision (Time for Assessment, 23[1]). SARA also authorizes any interested group or member of the public to petition

COSEWIC for a listing decision (Applications, 22[1])—a public participation clause that also plays a key role in the evolution and enforcement of ESA threatened or endangered species listing decisions. Furthermore, Canada's SARA law enforces on COSEWIC a 90-day deadline to make public its intention on how to list a certain species (Report on Response, 3).

A designated authority directed to assess and list species facing threats from hunting or human development, including urban sprawl, agriculture, or industry. Mandated public notification and an avenue to allow for and even encourage public participation in federal listing decisions. And timelines for actions. SARA also provides for critical habitat designations and spells out how Canadian authorities are authorized and directed to manage species and habitat in ways that ensure the long-term survival of both. The details vary, of course, and in many important ways. However, the bones are largely the same.

Canada even emulates the letter and spirit of the Endangered Species Act at the provincial level. The ESA's influence on endangered species management regimes is felt both nationally and subnationally, and in many ways locally. It truly is a revolutionary piece of legislation, as I keep arguing.

Manitoba, Canada, 1990 Endangered Species and Ecosystems Act

The provincial government of Manitoba adopted its 1990 Endangered Species and Ecosystems Act (ESEA) as an updated version of earlier endangered species management legislation. Much of it is familiar to experts in species conservation law.

With ESEA, the provincial lieutenant governor is authorized to appoint a minister in charge of threatened and endangered species policy direction and enforcement. The newer ESEA also establishes an advisory committee with members appointed by the lieutenant governor (Part II Administration, 6[2] and 6[3]). Upon receipt of a report by the advisory committee and minister, the lieutenant governor is authorized to determine whether a species can be listed as endangered, threatened, extirpated, or of "special concern" (Part III Species at Risk). After a listing determination, a recovery plan must then be drafted (Part III Species at Risk). A recovery plan may involve a special designation of specific habitats or areas of Manitoba deemed critical to the survival of a particular species, as well as legal restrictions on activities that might be found to be detrimental to a species' chances of survival in the wild (Part III Species at Risk). Here, Manitoba's law actually does one better than the U.S. ESA—the ESEA empowers Manitoba's lieutenant governor to even designate "endangered ecosystems," a provision obviously mimicking the ESA's authorization of critical habitat designations but stronger in tone and intention (Part III.1).

Public notification and participation are also required in Manitoba's law: the ESEA requires a 90-day public notification prior to a listing decision or enacting a new regulation to give time for the public to comment and request or recommend changes (Part III.1, 12.5[1]). And though drafters of listing decisions don't necessarily have to follow through on public comments, they often do (though they are free to ignore the public's comments and suggestions as long as regulators showed that the interests and concerns of involved members of the public have been given due consideration).

Again, we can see the same general pattern or framework emerging. An endangered species law designates a central authority. This authority is directed to investigate the status of species in the wild (within a certain jurisdiction) and then to make a determination about whether a species faces the prospect of future or imminent extinction. Listing decisions are made public to invite public participation, and a plan is enacted whereby the authority is empowered to protect not only animals and plants, but the very habitat and even the general ecosystem that sustains those animals and plants.

This general pattern isn't limited to the United States and Canada or even to the Western Hemisphere. The ESA's influence on other governments' endangered species management regimes echoes well beyond North America's shores. With a little digging, one easily can find echoes of the Endangered Species Act in species protection legislation drafted in Australia, Europe, Asia, and Africa.

Australia, 1992 Endangered Species Protection Act

Australia's 1992 Endangered Species Protection Act (ESPA) established the Endangered Species Advisory Committee under the direction and supervision of the National Parks and Wildlife Service (Part 1, 3[2][e]). The committee is directed and empowered to undertake assessments of species status in Australia's wild and then to make a determination about whether to list a species as endangered, vulnerable, or presumed extinct (Part 2, Listing, 14). The ESPA law goes into further detail, providing definitions for each listing category and how determinations are made according to each category. Listing decisions are made public in both a national Australian government periodical and newspapers circulating in states where a particular species is known to be found (Part 2, Division 2, 18). Yes, these laws were drafted when newspapers and not the internet were the primary vehicles of disseminating information to the broader public, but I'm assuming these sorts of clauses pertain to the newspapers' online presences, as well.

ESPA also invites members of the public to nominate species for certain listing statuses (Part 2, Division 2, 25). A species recovery plan also must be drafted by the responsible authority and made available for public review and comment (Part 3, Division 1). Australia's ESPA also details a timeline for action and even includes a detailed timetable for determining the precise steps to be taken and length of time authorized for implementing and reviewing recovery plans (Part 3, Division 2). The law also directs the ultimate regulatory authority, the minister of Australia's National Parks and Wildlife Service (equivalent to the director of the U.S. National Park Service or even the U.S. Secretary of the Interior) to consider and accept advice on any listing decisions offered from the Scientific Subcommittee. And, naturally, the Australian public must be notified of decisions on listings or delistings within 30 days of such a determination (Part 2, Division 2, 24). All of this is vaguely familiar, right?

Ireland, 1976 Wildlife Act

Ireland's Wildlife Act of 1976 also follows the broad pattern pioneered by America's ESA, though it was worded in a way that's far different and less recognizable than what I read out of Canadian and Australian endangered species law.

Ireland's law established the Minister of Lands as the primary authority responsible for species management and protection, again for the benefit of the citizenry (Part 2, Chapter 1, 11-1). The law directs the minister to consult with other government agencies, a committee of experts, and the public in determining threatened or endangered species policy, including, of course, actions that should be taken to prevent species' extinctions (Part 2, Chapter 1, 12-1 and 13-1). The Wildlife Act directs Ireland's Minister of Lands to identify habitat of importance to protected species and to set up mechanisms to protect the habitat so that it may continue to sustain that species, including through establishing wildlife refuges and other protected lands (Part 2, Chapter 2). The public must also be notified of any actions or changes in policy, as outlined in several sections of the act, and within a specified period of time. The Wildlife Act may be phrased differently—and even slightly poetically in some instances—but the outline is more or less the same.

The Philippines, 2001 Republic Act No. 9147 (Wildlife Resources Conservation and Protection Act)

Endangered Species Act mimicry isn't the sole purview of Western countries, either. It's found in Asian contexts, as well, despite the sometimes starkly different cultural perceptions of the value of wildlife and wild plants between East and West.

The Philippines Republic Act No. 9147, enacted in 2001, follows much of the general pattern that ESA pioneered, starting with establishing or confirming a responsible enforcement authority. The law establishes the Philippines Department of Environment and Natural Resources as the primary authority charged with enforcing the act as it pertains to terrestrial plants, animals, and ecosystems. Meanwhile, the Philippines Department of Agriculture is vested with species protections and natural resources extraction enforcement authority in aquatic ecosystems, both oceanic and freshwater (Sec. 4). This division of labor closely resembles the U.S. Department of the Interior's decision to split enforcement of the ESA between the Fish & Wildlife Service for continental ecosystems and the National Marine Fisheries Service for oceanic ecosystems.

Republic Act No. 9147 directs the secretary of either responsible department to make a determination as to whether a species should be listed as critically endangered, endangered, vulnerable, or "other accepted categories" (Sec. 22). The act then directs the secretary to make public a list of species and their statuses. There's also a public participation component, as in the West—the act orders the two departments to accept petitions from the public, interested groups, or individuals concerning various government-proposed threatened or endangered species listing and delisting decisions (Sec. 22, d). The Philippines' endangered species law also stipulated that either respective secretary has one year from the initial enactment of the legislation itself to develop a comprehensive list of species and their statuses and then two years after initial enactment to determine what critical habitats beyond already protected areas must be protected to ensure the survival of listed endangered plants and animals (Sec. 25). The Philippines' endangered species law then directs the responsible authorities to formulate precise conservation plans in collaboration with national and local governments and to make these plans open to public scrutiny.

The remarkable legacy of the Endangered Species Act—the "template" it gave other national and regional governments—reverberates even today. If I were able to read other languages aside from English and Japanese, I'm willing to bet I would discover traces of the ESA in endangered species laws enacted in other parts of Asia and Latin America. I say this because its inspiration is evident in very recent species protection legislation passed in Africa, or at least in one African country where English is commonly used.

Uganda, 2019 Wildlife Act

Uganda's 2019 Wildlife Act (Act 17) sets out on a different foot. It begins by stipulating clearly and loudly that all wild fauna and flora in the country are forever the property of the national government, to be managed by the government, ostensibly for the benefit of the people of Uganda unless a specific wild plant or animal has been lawfully taken possession of by any individual or group (Sec. 3). The ESA commodifies nature in this sense, as well, but Uganda's Wildlife Act makes this conceptualization more explicit—all wild plants and animals in Uganda are either property of the government or property of individuals who've legally taken them, as the law asserts right away.

Things soften from there. Via the Wildlife Act, the minister (simply defined in the act as "the minister responsible for wildlife") is directed by law to make a listing determination on species existing in Uganda under the careful guidance of an advisory board, presumably a group of experts selected by the government qualified to undertake species assessments. The minister can then list a species according to an even broader set of categories: as extinct, extinct in the wild, critically endangered, endangered, vulnerable, threatened, nearly threatened, or data deficient.

Transparency is also required under Uganda's Wildlife Act. A listing determination must be made public in the *Gazette*, a government periodical akin to the *Federal Register* in the United States (Part V, Sec. 34, 3). The longer list of status options—extinct, extinct in the wild, critically endangered, endangered, vulnerable, threatened, nearly threatened, or data deficient—more or less parrots the classifications or categories used by the International Union for Conservation of Nature in its periodic global species assessments. So here Uganda isn't only following the path broken by the U.S. ESA, but clearly took cues from broader international civil society when it drafted its very recently passed Wildlife Act.

Uganda's law does not specify any length of time by which the minister must arrive at a listing determination, nor is there a specified time limit for the government to issue any public notification of a listing determination, at least, as near I can tell from my reading of the act. So here Uganda's endangered species legislation departs from common practice. The Wildlife Act also is much more focused on individual user rights and puts a fair amount of emphasis on what processes exist for an individual or group to acquire a license to exploit wild species, including for bioprospecting purposes (the practice of harvesting species for their potential genetic value). Again, the law makes it clear that Uganda's government and society see their wildlife and wild plants as property to be exploited in a fair and sustainable way. The ESA also includes provisions for licensing and permits for any "incidental takes" of species that may occur, including those deemed necessary or perhaps unavoidable in the pursuit of species conservation, but it makes no mention of bioprospecting. This, I believe, is a reflection of the Wildlife Act's youth and an indication that drafters of Uganda's act had more contemporary concerns in mind when finalizing their endangered species law. The Wildlife Act was passed while parties to the United Nations Convention on Biological Diversity were amid a heated debate over global bioprospecting rules.

There are differences, yes, including important ones. Every country takes a slightly different approach with its respective endangered species conservation legislation. My analysis found no explicit proof of outright plagiarism, or countries lazily copying whole sentences or paragraphs from the ESA. Still, even Uganda's 2019 law contains the same general core provisions found in other countries' endangered species laws. You will find within them language designating a national authority charged with undertaking species assessments (with expert input), making listing decisions, and then notifying the public of said decisions while allowing the public to both give input on decisions and offer their own listing addition or revision proposals.

America's Endangered Species Act inspired Japan's government, as well.

As with the U.S. ESA, Japan's 1992 Act on Conservation of Endangered Species of Wild Fauna and Flora (ACES) begins with a declaration of a purpose or reason for drafting legislation to prevent extinctions. The Japanese government's stated purpose for conserving species is identical to that first articulated in the ESA: because maintaining greater biodiversity is of benefit to humans.

Chapter 1, article 1 of ACES declares that Japan has drafted this legislation "in view of the fact that wild fauna and flora are not only important components of ecosystems but also serve an essential role in enriching the lives of human beings," and because biodiversity helps contribute to "wholesome and cultured lives for present and future generations of citizens." Though culturally distinct and often far apart in terms of attitudes toward living natural resources (an excellent example is the rift between the

United States and Japan on the question of commercial whaling), these lines show that Japan has also embraced an originally American perspective that humans should seek to prevent wild animal and wild plant extinctions so that the humans, and not the animals or plants in question, may experience greater joy and other benefits.

As in the ESA, Japan's ACES designates an ultimate authority responsible for endangered species designation and protection; in Japan's case, this is the Ministry of the Environment (MoE). Originally, ACES empowered the former Environmental Agency, and this agency was upgraded to the status of a formal cabinet ministry in 2001. The law gives MoE's chief, the Minister of Environment, the power and responsibility for declaring species to be "rare" plants or animals warranting formal government protections (Article 2, 1). MoE inherited this authority from prior Japanese law and practice.

The ACES law also empowers the Ministry of the Environment to make a unilateral temporary declaration of a species' "rare" status (Article 5, 1). The law deems the Minister of Environment the responsible authority for enforcing ACES rules. It also requires MoE to make public any decisions on listings or conservation measures in the national government's own *Gazette*. The law specifies that an initial listing determination is considered a temporary designation, good for only three years (Article 5, 3 and 4). The ACES law also specifies the kinds of restricted activities connected to a formally designated "rare" species that the minister is authorized to regulate and the types of penalties individuals face for noncompliance.

As with America's ESA, Japan's ACES law also empowers MoE to designate and protect habitats deemed critical for the survival of rare plants and animals (Article 36, 2). Before making such a call, MoE is directed to consult with other relevant government authorities and with a Central Environmental Council, a body established in prior chapters of the ACES law (Article 36, 4). Japan's ACES law also demands transparency and public accountability. MoE is required by law to notify the public of any forthcoming changes to species or habitat status designations prior to making such determinations. The Japanese public also is given the right to provide its input into MoE's endangered species management decision-making processes (Article 36, 5).

The language of ACES makes explicit the role of the public in species conservation. The law goes above and beyond ESA here, a difference that I think stems from the different histories of endangered species management in the United States and Japan. ACES authorizes the head of MoE to appoint individual members of the general public as "rare wildlife species conservation promoters," given their expertise or even just their enthusiasm for rare species conservation (Article 51, 1). ACES directs MoE to cooperate with the nation's zoos and botanical gardens on species conservation initiatives and captive breeding programs where they may exist (Chapter V: Certified Zoos and Botanical Gardens Conserving Rare Species). ACES also directs MoE to educate the public on the importance of rare plant and animal conservation (Article 53, 2).

As it's drafted, ACES navigates an interesting balance between government authority and private citizen participation. I think this is a consequence of the fact that, at least for some endangered species in Japan, especially in the case of the red-crowned crane, many conservation initiatives began as grassroots community efforts. The national government only came in later to assume control and better organize these initiatives. In other words, ACES makes MoE and its minister the ultimate authorities over rare species conservation decisions, but the law also requires the minister and MoE officials to take every opportunity to invite public participation and cooperation at nearly every step of the endangered species conservation process. It's a reflection of the government's recognition that oftentimes communities take the lead, with government following later.

This language in ACES is similar to how the ESA requires public notifications and comment periods, invites public petitions for listing decisions, and directs authorities to incorporate and cooperate with other levels of government (including foreign governments) and members of the public in the endangered species conservation process. But ACES appears to take the importance of public participation even further. By reading the law, one gets a sense that ACES's drafters were of the firm conviction that species protection in Japan would fail without complete public buy-in and public responsibility. America's ESA approaches things in a more top-down manner, possibly a reflection of the fact that in America it was the government that initially took the formal lead in preventing extinctions—think of the very case of the whooping crane.

Arguably, ACES differs most from the U.S. ESA in the way it goes into great detail concerning how MoE and the minister may authorize and regulate various business dealings concerning designated rare species. I think this is a reflection of the fact that the ACES law is also designed to clarify how Japan complies with its CITES obligations. Yes, the ESA also mentions CITES, but only in passing. More recent endangered species laws passed many years after the ESA go into more detail about how governments plan to meet CITES compliance and implement international endangered species trade restrictions.

ACES is also not as detailed as the ESA. What I mean by this is that ACES appears to forego reliance on different levels or categories of extinction threats (going with a more general "endangered" or "rare" definition and avoiding clear, more precise language such as "threatened" or "near threatened" species). Nevertheless, ACES borrows the same general framework (the bones, if you will) first codified under America's ESA law. As with the ESA, Japan's foundational endangered species law designates a specific central government authority responsible for species assessments and for determining species' status, in collaboration with specially designated experts and members of the public. This central authority determined by ACES, MoE and its minister, must also designate habitat areas deemed critically important for the survival of an endangered or "rare" species and then impose restrictions or guidance over what activities are allowed in these areas, all with the aim of conserving species and preventing their extinction in Japan.

ACES also regulates timelines. Decisions on status designations must be made public and in a timely manner; for example, ACES specifies 14 days as the minimum length of time MoE must give the public prior notice of any changes, additions, or subtractions to endangered species and special habitat designations (Article 36, 5). ACES establishes a supreme power over endangered species management, but the law also mandates transparency throughout the decision-making process and ample opportunities for both public comment and public participation in the conservation process, all concepts likely borrowed in some way from the U.S. Endangered Species Act.

After reading through ACES, both the original Japanese and an online translated version, I concluded that the nation's legislators obviously had done their ESA homework prior to rolling up their sleeves and codifying extinction prevention rules within Japan's own domestic laws and cultural tilts.

So what can we conclude from all the above? First, I need to clarify something: the existence of the Endangered Species Act is not the only reason why multiple nations felt compelled to draft their own similar national and subnational endangered species laws.

Many comprehensive national, provincial, and state endangered species laws reference the UN Convention on International Trade in Endangered Species of Wild Fauna and Flora. America is no exception here—mention of CITES shows up early in the text of the Endangered Species Act. The United States signed onto CITES the same year the ESA was drafted and finalized, in 1973, and the act devotes an entire section whereby the federal government appoints the Secretary of the Interior as the managing authority for U.S. compliance with CITES while outlining how the United States plans to pursue said compliance (Convention Implementation Sec. 8A).

Other governments followed this script, and in many cases, governments actually weighted their domestic endangered species laws more heavily toward CITES trade rules compliance than explicit domestic protections for endangered species. In two cases I found, endangered species laws focus almost entirely on managing endangered species trade restrictions rather than domestic conservation initiatives. This may be a reflection of the relatively small size of the particular territorial jurisdictions falling under these pieces of legislation. For example, Singapore's Endangered Species (Export and Import) Act of 2006 explicitly states that it exists for the purposes of clarifying and directing the government's compliance with its CITES obligations. Norfolk Island, once a self-governing territory of Australia, passed its own Endangered Species Act in 1980, but the Norfolk Island ESA was drafted entirely as a CITES compliance vehicle; that act even includes the entirety of the CITES treaty text in its schedule addendum. Other nations adopted endangered species laws that can perhaps best be described as hybrid approaches, incorporating elements of explicit CITES compliance language as well as language specifying domestic endangered species conservation programs and measures.

But in multiple examples of national, provincial, and state endangered species legislation, a clear pattern is apparent and one with obvious American roots.

The United States Endangered Species Act of 1973 states that it is the duty of the government to prevent species' extinctions because greater biodiversity is of benefit to humankind. The entire world echoed this stance in the preamble of the 1992 United Nations Convention on Biological Diversity when UN member nations adopted language certifying member states' awareness of "the intrinsic value of biodiversity and of the ecological, genetic, social, economic, scientific, educational, cultural, recreational, and aesthetic values" worthy of legal protection.

From there, nations' endangered species protection laws adhere to a basic script written first in the Endangered Species Act.

The ESA law identifies a governing authority responsible for national endangered species protections. It specifies that this authority can assess and identify endangered species according to certain standards that it determines but that these standards must be informed by science and scientific advisers. The ESA demands full transparency, ordering those determinations to be made public and the public be afforded sufficient time to consider proposals and to respond. The public is also empowered to petition for endangered species determinations or changes to species' statuses.

The approach, evident again and again in other endangered species laws, has endangered species protections and enforcement powers placed under a central responsible authority while mandating that this central power or responsible government agency must cooperate with the general public to achieve the broader goal of preventing species' extinctions. Because of this fact, endangered species laws empower the public to appeal strong central authorities for policy changes. And the entire process is made open, transparent, and participatory. The national endangered species laws of other English-speaking nations and Japan embraced this same foundational pattern, as near as I can tell. This, to me, is a clear demonstration of the degree to which America's 50-year-old Endangered Species Act set the tone for government-driven biodiversity protections throughout the world.

This is constructivism at work, I believe. On matters of biodiversity and species protections, the modern peoples running modern governments wish to be seen as good actors. In pursuing this, they take cues from each other regarding what this really means or looks like. The United States largely took the lead back in 1973. Other governments emulated this example, provided as they were with a convenient template to follow. A local solution to a global problem, in other words.

Could this same pattern work to help mitigate global warming? It's an interesting idea at least.

International environmental treaties either assign rights to resources or areas with resource potential, facilitate information-sharing regimes in the pursuit of natural resources management, or encourage nations to protect the natural world existing within their own territories by use of incentives (lots of carrots, very rarely any sticks). In a sense, over the years the U.S. Endangered Species Act has evolved into a means of encouraging environmental protections far beyond the borders of the United States itself, all without convening any diplomatic conferences, bypassing the United Nations system and processes entirely. It could be the ultimate display of soft power.

It isn't easy to get a nation to join any global environmental treaty. Most such treaties are rather weak or weakly enforced, like the UN Convention on Biological Diversity. Unfortunately, it has to be this way. Otherwise, most nations would decline the opportunity to sign on to multilateral environmental agreements, or MEAs, as academics and diplomatic circles are fond of calling them. Adherents to the realist school of international relations theory could perhaps best explain to us why this is the case.

Circumstances vary, of course, but generally speaking, to get any sovereign, independent nation-state to sign on to your global environmental protection treaty, you have to promise them essentially three conditions.

First, you must ensure that hypothetical government that its entirely voluntary participation in your global environmental treaty is relatively inexpensive—that it won't cost it too much to help maintain a secretariat, to dispatch representatives to meetings, or to pay for domestic measures necessary to participate. For some nations, perhaps most, participation in the new treaty body may even be entirely free of charge. I can understand why. Governments prefer to spend their money on their own citizens first. They seek to avoid additional budgetary expenses, not add to them (generally speaking, of course). So oftentimes an international environmental treaty is crafted in a way that makes use of existing resources or capabilities or recruits from richer nations the funds necessary for helping poorer participant nations build up new capacities. As long as it doesn't cost them too much money, that government may be more than happy to join your treaty.

Second, your amazing super-duper international world ecosystem protection and defense treaty best not actually require all that much of its signatories. In other words, any additional burdens placed on nations for their participation should be kept to a bare minimum if at all possible. By doing this, you ensure that the new treaty body doesn't tread on the preexisting domestic affairs of any would-be member state too much. To further increase the odds that governments sign on to your treaty, the language in the treaty itself should make it clear that nations are left to be as active or inactive in this new global environmental protection architecture or treaty body as it suits them.

Third, make sure that failure to comply with any or even all of the provisions of your new treaty will not result in any fines levied or punishments rendered to the noncompliant party. There must be no real-world penalties or consequences for nations either slipping a bit in their compliance efforts or even ignoring their responsibilities under this new treaty entirely. Noncompliance should definitely be consequence free, otherwise governments may hesitate to join.

Those are the three conditions needed. This is coming from someone who has spent a very, very long time watching negotiations at the United Nations and the way diplomacy actually works there. It is almost beyond the theoretical at this point (not entirely, of course) but firmly grounded in the practical and pragmatic. One, make participation in your international eco-treaty inexpensive or free. Two, make the burden minimal. And three, make actual compliance optional and noncompliance consequence-free. Meet these basic conditions (and several others as well, of course) and you've got the makings of a well-subscribed international environmental protection treaty. As I hinted at earlier, the UN Convention on Biological Diversity meets all these preconditions. No wonder 196 nations have signed on to that MEA.

But this describes a top-down approach to global environmental management. I would argue that the U.S. Endangered Species Act introduced a unique, new, and highly effective bottom-up mechanism. Let's call it "inspired mimicry" or "benevolent mimicry" for lack of a better term. It's an interesting and rare approach to international affairs that I believe is very much rooted in the constructivist school of thought and shows the true strength of soft power.

America took the initiative to craft a thorough law regarding extinction prevention and endangered species management. In doing so, America in a sense demonstrated to the world how a responsible state acts in the face of a worsening extinction crisis. Seeking to become good actors in their own rights, other states followed, giving rise to the endangered species protection architecture that we see today, as I've described it above.

You may recall, perhaps, that earlier in this book I questioned the popular premise that global problems necessarily require global solutions. I've heard this phrase repeated so many times at the United Nations and beyond that I wish I could trademark it and charge 10 cents every time it's used. Diplomats and world leaders love saying this for virtually every world-spanning problem—global problems require global solutions, they say. They, of course, say this for the global extinction or biodiversity loss crisis as well. It is no exception here, they insist. Only this statement isn't true—global problems don't always require global solutions. This statement doesn't even hold logically, if you really think about it.

The global extinction crisis was absolutely roiling in 1973 when the Endangered Species Act was first passed. Still, the U.S. Congress didn't wait for the world or the member states of the United Nations to unite and come forth with a common overarching global plan to address it. They focused only on the problem as it pertained to the United States and proceeded to craft one possible solution. Rather than wait for collective action, America simply moved on the issue alone, drafting, debating, and eventually enacting a massive species conservation law that's only enforceable within U.S. jurisdictions.

This was a local solution to a global problem. Since then, the ESA has had a tremendous impact on how other nations both perceived the problem and imagined potential solutions to it. Dozens of nations followed America's steps and came up with their own local solutions to this global dilemma, inspired by the ESA. Thanks to this benevolent and entirely benign mimicry, thousands of animals and plants that were once on the brink of annihilation in Africa, Asia, Australia, Europe, and South America are still happily and thankfully with us today. Perhaps we can achieve something similar today, rather than wait for the UN family to collectively halt the ongoing extinction crisis.

A local solution to a global problem—this *is* one option, folks.

Here's some more food for thought.

The world has clearly drawn lessons or inspiration from the way America legislates and pursues endangered species protections. Perhaps it's time for that learning to move in the other direction. Perhaps America can learn a thing or two from other countries' experiences.

The whooping crane has been saved from the brink of extinction. The recovery of the red-crowned crane of Hokkaido has been much more impressive. Though found on opposite ends of the planet, both species are very similar and clearly derived from the same ancestor. The two separate conservation histories behind these two species parallel one another in interesting ways. It wouldn't be unexpected that their rates of recovery would also more or less match. But that's not the case.

From my observations of and investigations into these two case studies, there seems a strong case to be made that the artificial feeding campaign in Hokkaido resulted in higher rates of reproduction and much faster population growth, making it possible for the red-crowned crane to be removed from endangered species lists far sooner. U.S. Fish & Wildlife may desire to do the same for the whooping crane, but it's clear to me that it is far too early to do so, and I am confident that many others agree and will thwart any attempts to delist or change the status of the whooping crane in the courts—after all, they've succeeded here for the gray wolf for years. But what if the Fish & Wildlife Service had introduced and then sustained, over many decades, a winter feeding program for the whooping crane, or even a drought intervention feeding program? Would that species' numbers look different today? It's a compelling question.

There are important cultural distinctions that can be made between the way Japan and the United States tackle the extinction crisis, as well. And these different cultural approaches to endangered species management are evident in the two nations' separate endangered species laws.

Reading ACES, I can't help but notice a bottom-up framework emerging. We often think of Japan as a very top-down, heavily regimented society where the government takes a prominent role in everything, and this is true for the most part. But ACES stands out from the ESA in its emphasis on the importance of community engagement and reliance on members of the public to further the mission of wildlife conservation. Though ACES clearly draws inspiration from the Endangered Species Act, the ESA reads more like a top-down approach (with public participation mandated, of course), whereas parts of ACES read as if the conservation ethos in Japan had emerged in a bottom-up fashion. The tale of these two cranes could help to explain how this interesting contrast came to be.

Whooping crane conservation has been a government-led initiative for the vast majority of its history. Nongovernmental organizations of course play an important and influential role and one the U.S. government recognizes. But saving the whooping crane from extinction has been first and foremost a federal government prerogative paid for with federal dollars and enforced with federal laws. This reality may explain why the Endangered Species Act begins by clearly delegating which government authority shall yield the power of the act itself.

Efforts to save the red-crowned crane began as community-led initiatives centered almost exclusively on a couple of small agricultural villages in eastern Hokkaido. Members of these communities set aside open spaces and launched artificial feeding programs without any encouragement or directives. They did this purely out of concern for the survival of *tancho*, assumed to have vanished from Japan a few decades prior. Japanese government agents only later formalized these unique arrangements. Perhaps this explains ACES's language on identifying and officially recognizing interested and engaged community members. The law calls on the central government to deputize members of the public as agents for the state in the name of species protections. I think this history helps to unveil why ACES is written the way it is.

This tale of two cranes is also a tale of two endangered species protection laws, to some degree at least. The Endangered Species Act is far older, of course, now entering its golden 50s. There are numerous lessons to be learned from this 50-year history, but these lessons don't emanate only from the Endangered Species Act itself.

We Americans are often loath to hear this, but it needs to be said: there's much to be learned from the way things are done in other countries. We don't have all the answers.

CHAPTER SEVEN

Cautionary Tales

Here's a provocative question: does America's Endangered Species Act actually work?

In other words, does it achieve what it sets out to achieve? Will listing a species as threatened or endangered under the ESA improve the chances of that species surviving and perhaps even thriving in the future?

Most observers, I believe, would argue that, yes, the ESA does work. To many of them, this would seem rather obvious. After all, there are plenty of fantastic examples to support this assertion. The whooping crane is possibly one of the most famous ESA success stories. I believe that goal of 1,000 AWB whoopers eventually summering in Wood Buffalo National Park will be achieved, and likely not too long from now. It will be a remarkable achievement for conservationists, considering that the worldwide whooping crane population was once so low you could fit every single living whooper into a bus.

Most studies generally agree with the view that the ESA works—research has found that listing a species correlates with that species' survival and recovery. The ESA doesn't always succeed, of course (and the law's more ardent proponents openly admit this), but by and large the data seems to show that protecting animals and plants under the ESA improves their chances of survival. The law saves animals and plants from extinction. But not everyone is so willing to instantly declare the ESA a hands-down runaway success story.

About a decade ago, two Canadian researchers took issue with this popular view. They argued that the more you drill down, the more you find that the ESA isn't particularly effective after all. It's a contrarian and potentially controversial take, which is why I'm naturally attracted to it, though I don't necessarily agree with their views (another point lost in all the mudslinging that makes up modern media discourse and politics: it is absolutely possible to find ideas or arguments interesting or compelling without actually agreeing with them). How did these two Canadian researchers come to this rather startling conclusion? Let's take a quick look.

Katherine Gibbs and David Currie of the University of Ottawa said they took note of the multiple peer-reviewed and published research in the academic literature that found the ESA to be an effective tool for species conservation. Reading through their paper, I get the impression that they were left rather unimpressed. What seems to bother them most is that none of these earlier studies sought to determine the relative degree to which the ESA's main management tools are effective or influential. If you believe the ESA is effective, then there must be a way to quantify its effectiveness, no? That seems to have been their thinking at the time.

In their view, the three main tools the law brings to the table in the battle to prevent senseless extinctions are the funding that the ESA opens up for conservation initiatives (authorizing spending on land acquisition, maintenance, research, etc.), the species-specific recovery plans mandated by the law, and the law's stipulation that the Department of the Interior identify and designate "critical habitat" that officials must preserve and protect to ensure the survival of a given endangered species of plant or animal. So Gibbs and Currie focused on these functions in their study.

The researchers also wanted to see if there was any clear correlation or apparent causality between the length of time a species had been listed as threatened or endangered and population recovery. Earlier researchers found such a correlation, but Currie and Gibbs thought it important to determine whether the links are meaningful or not. "Earlier studies sometimes found that statistically significant effects of these tools could be detected," the researchers wrote, "but they have not answered the question of whether the effects were large enough to be biologically meaningful."[1]

Gibbs and Currie pulled the Department of the Interior's official species status reports to Congress for 1,179 species, analyzing reports dating from 1988 to 2006. They cross-referenced these details to data and statistics on conservation program funding levels, the number of years species were on the endangered species list (for species so listed prior to 2003), the number of years conservation plans had been in place, and the length of time of a critical habitat designation, along with selected peer-reviewed research on a particular species. They mostly looked at plants, which comprised about 61 percent of their sample size, 9 percent of the species they zeroed in on were fish (it's a little-known fact that most listed threatened or endangered vertebrates are fish), and only 6 percent of their sample size was comprised of bird species.

Their conclusion: where's the beef?

Though Gibbs and Currie said that they thoroughly reviewed "more species, more indicators of recovery, and more variables that potentially influence recovery than any earlier study," they were left deflated. "We still find only weak effects, or none at all," they wrote.[2]

It wasn't all bad news, to be clear. The researchers found some differences depending on the category of the sample. For instance, they found the number of years listed and recovery success for bird species to be "significantly positively" correlated. Recovery for mammals in general tended to correlate to the amount of conservation funds spent on that recovery. But overall, the effects of listing time and funds spent were weak at best, they reported. "The best among the weak predictors of recovery in our study is the number of years a species had been listed," Gibbs and Currie said, throwing ESA proponents a bone. Two possible conclusions can be reached here, they argue: either the benefits derived from the ESA during the 18 years considered in their review are modest at best, or there simply isn't enough data or data points available to draw any definitive conclusions yet, let alone any broad-reaching conclusion that a surly member of Congress might grasp for.

But there's a caveat.

Gibbs and Currie are careful to say that, regardless of how one interprets their results, their findings do not mean that the U.S. Endangered Species Act has no utility or function whatsoever. They acknowledged that the law has prevented extinctions and helped many endangered species to recover in population sizes. This is undoubtedly true, and I believe the rebound of the North American whooping crane is proof of this. The researchers are talking about averages here. They note that previous studies said they detected a connection or correlation between endangered species recovery success and funding, management planning, and habitat designations. But these prior studies didn't try to measure the relative strength of any existing effects or influence. Gibbs and Currie reached the conclusion that, on average, any effects discerned are either weak or barely existent.

What do you think?

For me personally, I think the main problem with the Gibbs-Currie study is the timeline they looked at: congressional reporting during an 18-year period. For some species, this may be more than ample time to detect policy effects on recoveries, I suppose. For others—perhaps the majority of listed endangered species—I'm willing to bet this isn't nearly enough time to get a clear handle on things.

Endangered species recovery work can be an incredibly drawn-out, time-consuming process. To their credit, Gibbs and Currie acknowledged this, mentioning the possibility that their review may simply be data insufficient. But I understand the instinct to look into these things. Money, legal action, personnel decisions, and more are involved in the policy decisions and actions undergirding the ESA. Congress obviously wants to know. But looking only at averages or even the entire swath of listed species, then it's probably still too early to say whether the Endangered Species Act is achieving its objectives, even though the law has been in force for some 50 odd years now. The Endangered Species Act has many more years, decades, perhaps even centuries to play out; 18 years is almost a blink of an eye in the grand scheme of things.

Others have argued that the Endangered Species Act has been proven to be a highly effective instrument, and some have even sought to quantify its effectiveness.

For instance, in a 2005 assessment in the journal *BioScience*, a team of researchers made the same point as just I made: weighing the success of the ESA over a relatively short time frame may be a fool's errand. Instead, it's far better, those researchers argued, to take a long-term view, considering not only a lengthier time horizon going back in the past but also trying to peer out into the future a bit, to see whether the legislation and a listing improves the odds of a species' long-term survival, if this can even be done. In other words, "although few threatened and endangered species have fully recovered, the short time most have been protected . . . renders this a weak test of the effectiveness of the Endangered Species Act," Taylor and colleagues wrote. "A better measure is the extent to which the provisions of the ESA are moving species toward recovery."[3]

Taylor and his colleagues largely looked at the same data sources that the Canadian critics relied on: biennial reports to Congress on the status of listed species by the U.S. Fish & Wildlife Service and the National Marine Fisheries Service. The researchers don't characterize their findings resulting from this data review exercise as "weak" at all. "The longer a species is listed and subject to the regulation of take, the more likely it is to be improving and the less likely to be declining, irrespective of recovery plans and critical habitat," they wrote. "This suggests that imperiled species should be listed under the ESA as soon as possible." One problem they note, however, is the inadequate funding directed toward species conservation, at least back when they were conducting their research. List an endangered species quickly, then adequately fund a conservation plan, and you should see good results, they argued. That especially holds true for spending on habitat protections or habitat set aside. "Critical habitat was strongly negatively associated with declining trends in the early period and positively associated with improving trends in the late period, suggesting that it has been effective in assisting species recovery, despite administrative barriers," the research team wrote.[4]

Is there room for improvement? Of course there is. In a 2016 edition of the journal *Issues in Ecology*, Evans and colleagues point out several areas where the ESA is lacking, while still vociferously defending the act and its legacy in the United States.

Among other issues, they mention one key point that has not gone unnoticed: ESA protections focus heavily on iconic species like the whooping crane. The larger and more beautiful an endangered species is, the more likely it attracts attention and protection funding. Smaller, uglier things fare the worst, especially plants and insects. And most species protection funding that is mobilized ostensibly to abide by the ESA in reality benefits sportfishing—at least as late as 2016, most U.S. federal dollars devoted to endangered species protections were spent on salmon and sturgeon, as Evans and colleagues courageously pointed out (following its initial passage in late 1973, Congress later amended the ESA in a way that directed funds for fish breeding and creek stocking, mainly with anglers in mind).[5]

Endangered species management plans in the United States also tend to focus on managing but not eliminating long-term threats to a species' survival, suggesting that many animals and plants listed under the ESA may be fated to remain on the endangered or threatened species list indefinitely.

But it's a delicate dance, and we often don't know what effect any given management strategy might have until after the fact. Things often don't turn out the way we expect. In fact, sometimes the unexpected should be expected. Sometimes a management strategy can backfire, helping one species while harming another, even if the action is meant to restore an entire ecosystem to its original natural function.

Case in point, the California ridgeway rail.

This species of rail, a small coastal wetland bird, once thrived in the San Francisco Bay area. Then people came and paved most of the bay. What's left is a series of small patches of surviving wetlands, much of it overrun with invasive vegetation, in particular *Spartina alterniflora*, or smooth cordgrass. *Spartina* is a native of the East Coast, where it's in a seemingly never-ending battle with phragmites, an invader from Europe that grows into a densely packed, almost wall-like structure of vegetation. On the East Coast, *Spartina* barely holds its own, but it's overrunning much of the West Coast's bays and inlets. *Spartina* is considered the good guy on the East Coast—it adds to coastal shorelines while creating vital natural fish habitat and nurseries, much like mangrove forests do. On the West Coast and in California, however, *Spartina alterniflora* is deemed the enemy—it can disrupt natural water flow patterns in bays and impede coastal navigation. It must be bad for native species like the ridgeway rail, too, or so many officials there once thought.

From 2005 to 2012 or so, a team of conservationists decided to fight back. They launched a *Spartina* removal campaign in San Francisco Bay. In some places, they removed almost all the invasive *Spartina* that they could get their hands on. In others, only partial removal occurred. A noble act. But as Casazza and colleagues explained in the March 2016 issue of the journal *Ecology and Society*, this ecosystem restoration campaign ended up harming the California ridgeway rail, considered a threatened species at the time.

As invasive *Spartina* cover declined, so did the number of ridgeway rails inhabiting the coastal wetlands subject to this invasive plant eradication effort. Researchers estimate that rail populations declined on average by 9 percent per year throughout the entire region as the East Coast *Spartina* grasses were being ripped out; population declines were even more pronounced in those patches where the conservationists had undertaken the most vigorous *Spartina* removal.[6]

This ecosystem restoration effort in the bay was launched without considering the changed realities there. The real cause of the California ridgeway rail's decline wasn't the removal of *Spartina alterniflora*. It was the 200 years of human development, first agricultural and then urban. The invasive *Spartina* species actually created a unique niche for the California ridgeway rail. This invasive "enemy" had in fact become an ally as far as the rails were concerned.

The resident ridgeway rails learned to use the grasses to hide from predators, while finding delicious things to snack on in its cover. Ecosystem managers forged ahead with *Spartina* eradication anyway, without considering these changed circumstances. Here was a case where an invasive plant, normally detrimental to natural ecosystem functions and services, instead provided an ecosystem service of its own—saving the California ridgeway rail from extinction. Throughout the history of San Francisco Bay's development, 80 percent of the rail's original habitat was lost. The invasion of non-native *Spartina* seems to have slightly enhanced the utility of the small amount of habitat remaining, likely contributing to a slight increase in rail population. But not by much—the rail's regional population declined to fewer than 1,000 or so individuals until it was halted with the introduction of habitat restoration programs and predator controls. The invasive *Spartina* likely had a minor secondary effect, or so Casazza and colleagues concluded, but it was an effect, nevertheless, possibly saving this species. Then conservation managers came in and starting removing this one minor saving grace. Oops.

So what's the lesson to be learned here? Should we let the *Spartina* thrive unimpeded? I don't believe so. The answer isn't necessarily to let the invasive plant thrive (as it may threaten other species), but if we rip out the *Spartina*, we should compensate for the loss of the unique niche it creates. If you take out this plant, immediately replace it with native vegetation that can serve the same ecosystem function for the California ridgeway rail.

It's a tough lesson to learn—in some cases, endangered species habitat restoration efforts actually can end up harming, and not helping, an endangered species. We just need to make certain we understand what those circumstances are before forging ahead with a habitat restoration initiative. This is an increasingly recognized problem in ecosystem management and restoration initiatives. One excellent example is the encroachment of salt cedar up and down coastal waterways. Invasive plants and animals, over time, find their own niche functions, and some native species learn to rely on these invaders even as the invasive species pushes out or otherwise damages native fauna and flora. Clear management goals are important, but so is a clear understanding of precisely what's going on.

But the California ridgeway rail's story is just one case study.

Under most circumstances, invasive species are a significant challenge. Usually, the goal is to simply manage these challenges. Can we eliminate them entirely? Is this even possible? The lionfish is now found nearly everywhere in the Gulf of Mexico, Caribbean Sea, and Atlantic Seaboard. I have a feeling this Pacific invader isn't going anywhere. Perhaps the best we can hope for is steady and consistent management, at least until the ecosystem adjusts to the changed reality. Perhaps over time the circumstances will change, a different balance will set in, and the lionfish will find its own niche ecosystem function in its new home ranges (ideally as food for predators that can keep lionfish numbers under control).

Of course, then there's the ever-looming threat of global warming.

The rising global temperatures brought by climate change are causing shifts in the normal climatic patterns of ecosystems and biomes. "Biome" is a broad concept that encompasses multiple ecosystems—tropical forests, temperate forests, shrublands, grasslands, and deserts are all considered biomes. These changes are relentless and will continue for some time, meaning what was once a grassland could steadily shift to desert. Meanwhile, endangered species inhabit these shifting landscapes. Can these species adapt in time? Can they relocate to more hospitable climes and habitats? Barring adaptation or relocation, there's only one other option left: extinction.

But the greatest threat to wild species in the United States is the ongoing expansion of the U.S. population and the urban sprawl it causes, especially in the West. Although some species have likely succumbed to climate change, most competent studies show that the biggest causes of extinctions in modern times are human encroachment and associated habit loss. We know that species fare better in larger areas or in wider expanses of habitat and that species existing on islands are particularly vulnerable to extinction. By carving up the continental landscapes with bulldozers and plows and replacing these once wild expanses with asphalt and crops, we have essentially busted up America (and everywhere else) into millions of islands, some of them connected only loosely but most cut off from one another entirely. This "islandization" of continental landmasses is the greatest long-term threat to biodiversity in the world. But that's no reason to completely despair.

In the United States, the ESA is both powerful enough and flexible enough to allow wildlife managers to address persistent problems in species conservation and to head off future emerging threats to endangered species' survival. "The ESA has worked remarkably well to shield hundreds of species in the United States from extinction, and it remains one of the country's strongest environmental laws," as Evans and his colleagues said. "It is also a flexible statute that permits considerable *innovation* in its implementation."[7] My emphasis.

And as I attempted to demonstrate earlier, the ESA has emerged to become a powerful force in international species conservation efforts. It helped inspire species protection laws in other jurisdictions, both local and national, as foreign governments drafted and implement their own endangered species statutes by using more or less the same playbook Congress used when codifying the ESA into existence.

The ESA forged the mold and set the tone for the ultimate drafting of the United Nations Convention on Biological Diversity—the CBD is, admittedly, a very weak multilateral environmental agreement, but one that nevertheless has merit from a constructivist point of view. As with the whooping crane—and, some would argue, the red-crowned crane as well—it's probably safe to say that thousands of species of remarkable plants and animals that were nearly wiped out completely are happily still in existence today thanks to the U.S. Endangered Species Act, both in the United States and throughout the world. The act inspired the world. Many species haven't been fortunate enough to benefit from this local solution to a very real global problem. The vaquita of Mexico's Sea of Cortez may be next in line for extinction despite the global conservation mood and atmosphere of ESA-inspired legislation.

Fifty years on, what have we learned?

That the Endangered Species Act works. But not all the time. Still, there's plenty of time to make adjustments to ESA implementation as new information comes in. The act has been amended before, and it may be amended again, but already it provides conservation authorities in the United States with plenty of tools to turn back the clock on our current extinction crisis. The ESA also offers inspiration for foreign jurisdictions to draft their own laws or ordinances in service of the war against the Anthropocene mass extinction event now underway.

Moving forward, many suggested tweaks or fixes to the ESA and to endangered species management practice in general will no doubt generate plenty of heated debate and prove highly controversial. I'll start: perhaps it's time that we begin regularly feeding the wildlife, at least in some circumstances.

"*This* is the kind of reporting I want to be doing full time," I told my wife, notebook in hand as I jotted down as much detail as I could about the action that was happening right before our eyes.

My wife is a photographer. I brought her along that day so that we could finally get some great up-close shots of one of North America's most critically endangered species, the Attwater's prairie chicken. We'd been trying to spot one of these ground-dwelling but not flightless birds for a couple years now, and we failed every single time. But this time was different. This time, I pulled my reporter's card and arranged for a guided tour of one of the final refuges of this elusive species, once ubiquitous throughout the western Gulf of Mexico coastal region. It was a desperate move, but there was an important story to tell here. I wasn't too proud to beg.

The Attwater's prairie chicken's home is easy to find. Drive about 50 miles straight west from Houston, then head south of Interstate 10 around the town of Sealy for another ten or so miles. You eventually come across a simple blue sign inviting you to turn off onto a narrow dirt county road to see a bit of rare federal real estate in the heart of Texas (which is 95 percent privately owned, I should mention again). Nestled in the table-flat landscape, surrounded by nothing but farms and the occasional pleasant meandering creek, is the U.S. Fish & Wildlife Services' Attwater Prairie Chicken National Wildlife Refuge.

You cross a cattle guard to enter, but the refuge is noticeably different from the land that surrounds it. This landscape is devoid of crops or any signs of recent tillage, though much of it probably was once farmland. There's some oil and gas infrastructure still there (the federal government allows drilling and hydrocarbon extraction on its wildlife refuges; the Aransas National Wildlife Refuge hosts oil industry infrastructure, too). But the refuge is mostly wide swaths of tall brown prairie grasses broken up by a few dirt roads. Spring brings with it temporary patches of wildflowers and green leafy plants, adding colorful hues to the flat terrain, but most of this coastal prairie land dries up as the summer heat rolls in. The prairies of Attwater are dry and brown for much of the year.

That's what makes Attwater's prairie chickens so hard to spot—they're the color of the prairie grasses. They also know how to stay really still and well hidden, especially at the slightest hint of movement or approaching danger. The best time to spot these birds is during spring when males seek higher and less vegetated ground early in the morning, just past sunrise. It's at these little mounds, or leks, where they let loose the chicken dance, a critical mating ritual that had ensured the survival of Attwater's prairie chicken for eons before European settlers arrived to the Gulf of Mexico's shores.

The dance begins with a furious stomping of both feet, sort of like the prairie chicken is running in place. Do you remember that scene in the movie *Flashdance* when Jennifer Beals's character rehearses to the song "Maniac" by quickly high-stepping or pumping her legs up and down? It kind of looks like that. These dance moves are followed up with the male prairie chicken puffing up his colorful throat like a balloon and letting out a low-pitched "woo-woo-ooo" sound. Wildlife experts call this entire spectacle "booming." I call it the chicken dance. A male Attwater's prairie chicken has to know how to chicken dance if he's to have any hope of landing a date on a Saturday night.

Lady Attwater's prairie chickens love the chicken dance. Done right, a male prairie chicken can win the affections of any comely female that might happen to be nearby, thus winning the opportunity to sire offspring and give rise to the next generation of Attwater's prairie chickens. For reasons not entirely understood by me, the males of several species of birds have evolved to either sport colorful gaudy plumage or undertake elaborate displays of eccentric behavior to attract mates and win the chance to reproduce. This tends to make male birds stand out like sore thumbs, leaving them vulnerable to predation. But it must be worth it.

The first time my wife and I visited this wildlife refuge, we left early, before the sun was out, and made it just in time for the gates to open and to hopefully spot a male prairie chicken atop a lek booming away, trying to catch the attention of the few remaining females surviving there. We drove and drove through the flat grassy terrain. We stopped the car on occasion to stretch, walk around quietly a bit, and scan the horizon for any signs of chicken dance action. We didn't see a single Attwater's prairie chicken that day. Not one. Dejected, we stopped by the visitor center for a rest before heading back home.

On our second visit, we sought professional help—we opted to join one of the van tours the refuge's operators offer to visitors, reservations required, of course. Our driver/guide gave us plenty of great information and insight about the refuge—its history, the other species found there, talk of possibly expanding it—but he couldn't compel the chickens to come out to pose for photographs. We didn't see a single Attwater's prairie chicken during our second visit, either. Not even a footprint. To be fair, the Fish & Wildlife Service (FWS) warned us that this might be the case, and our kind and very knowledgeable driver/guide reiterated this point right before we set out that day, just to make sure his guests weren't expecting too much. It was like a whale watching tour—they can take you out on a boat to where the whales might be, but that doesn't mean the whales actually will be there. By this time, I had grown used to the disappointment. I paid for a whale watching tour in New Jersey once, too, and of course didn't see a single whale (though we did spot some dolphins that day).

It's entirely plausible for one to visit the Attwater Prairie Chicken National Wildlife Refuge multiple times, for several hours each visit, and never even set eyes on one single member of this refuge's namesake, ever. The entire refuge spreads across some 10,000 acres, or about 4,000 hectares. It's not a massive refuge, but as national refuges go, it's expansive and yet populated with so few of these critically endangered birds. When we paid our visits, Attwater's prairie chickens living there numbered only in the dozens. Like the whooping crane, the Attwater's prairie chicken came very close to vanishing entirely.

The U.S. government has been struggling to keep this species alive since the late 1960s, when Attwater's prairie chicken (also sometimes called the Attwater prairie chicken) became one of the first formally listed endangered species, alongside the whooping crane. Federal officials have been spinning their wheels on Attwater's prairie chicken recovery ever since, seeing this species' numbers rise and fall and then rise and fall again, never quite cracking any critical threshold that might indicate sustained recovery.

A second population of Attwater's prairie chickens is maintained at a private ranch near the community of Goliad further south, and there is an even smaller group on the grounds of Johnson Space Center in Houston. These places, along with the federal refuge near the community of Eagle Lake, are the only places on Earth where this species can be spotted in the wild, maybe (if you have better luck than we did). The rest are housed in a captive breeding program at the Houston Zoo.

When I lived in Texas, you could take every Attwater's prairie chicken in existence—including that captive population at the Houston Zoo—and fit them into a reasonably sized cargo van. Attwater's prairie chickens once numbered in the millions, the population spanning the entirety of the Texas and Louisiana Gulf Coast prairie, a unique and distinct ecosystem that was torn apart to make room for farms, ranches, roads, and cities. The prairie chicken's habitat needs were ignored entirely.

Just to illustrate the federal government's struggle with this bird: by 2011, an estimated 30 individual Attwater's prairie chickens inhabited the refuge. The species was then laid low by a nasty drought that occurred that same year, a parching that forced several Texas communities to implement water use restrictions. As I mentioned earlier, at the time, it was the worst single-year drought on record in Texas's history. The mean 2011 Texas drought was an event significant enough that I was eventually dispatched to central and west Texas for a reporting tour focused on global warming. I remember visiting the dusty and sweltering community of Llano and its nearly dry reservoir. Later, the rains returned and recovery eventually came to Texas; the prairie chickens' numbers rose to some 126 by 2016. Then torrential rains and flooding knocked the population back to just 42. The refuge's prairie chicken population started to steadily recover again, and then came Hurricane Harvey in 2017.

While I was busy bailing water out of my swimming pool in the pouring rain in a panicked and desperate attempt to keep my house from flooding, Hurricane Harvey was busy killing prairie chickens at the national refuge and the private Goliad ranch. Many of the poor creatures died of exposure, drenched wet for days and unable to find a warm, dry spot at night. The greatest single rainfall event in U.S. history came very close to rendering the Attwater's prairie chicken extinct in the wild. A spring 2018 count turned up just 12 surviving adults. This recovery effort has yo-yoed throughout its entire history. John Magera, manager of the refuge during my visit, said the 2011 drought was especially painful since he and other colleagues at FWS were growing confident that the Attwater's prairie chicken population was on the cusp of taking off. "That really hammered us," he told me.

On our third visit to the Attwater's refuge, Magera kindly arranged for us to ride along with FWS to check up on prairie chickens under their care. This is the type of reporting I wanted to do more of, I would later tell my wife. We were led into the field with wildlife conservation experts to witness firsthand interventions for the survival of one of the rarest and most endangered species in North America.

At the time, the Attwater's prairie chicken recovery effort involved a three-step process. First, Attwater's prairie chickens were bred and partly raised in captivity at the Houston Zoo. I saw this operation up close too; the prairie chickens are housed in coops on the zoo's property, right in the middle of the city, with veterinary medical facilities nearby for health checks as needed. Next, young chickens are driven to the refuge and placed in large cages set up in the field. This is to acclimate them to their new environment, I was told. The cages protect the chickens from birds of prey (the cages had to be reinforced for this purpose, too—Magera shared that he and his crew came out one day to find a prairie chicken missing its head). Finally, when the young adults are deemed old enough and ready for the big scary world outside their protective pens, the prairie chickens are radio-collared and then let loose into the refuge to fend for themselves and hopefully survive. After that, officials more or less cross their fingers and hope for the best, tracking them, of course, but otherwise following a hands-off policy, since they felt the refuge had plenty of food for the birds to enjoy.

We drove in Magera's truck to a spot where there were guaranteed to be Attwater's prairie chickens. After all, they were penned in and couldn't escape. I was growing excited. *Finally*, I thought to myself, *I can see one of these birds up close*. Sure, I was kind of cheating, but it didn't feel like it.

I've seen wild pheasants. The Attwater's prairie chicken is no pheasant. For starters, it lacks the long tail feathers of a pheasant. But it doesn't look like a chicken, either, although it is about the same size and stature. Its mottled plumage of brown, dark brown, gray, and maybe even hints of orange are excellent camouflage, perfect for blending in with the prairie grasses. Perhaps I had stumbled upon a free-roaming prairie chicken on an earlier visit to the refuge and didn't even notice. Later in life, the males take on showier plumage—in particular, their distinct, bright orange neck air sacks that they inflate while performing the chicken dance in an attempt to impress the ladies. Female Attwater's prairie chickens are rather plain in appearance but still remarkable in their own right.

Magera crawled into the pen and grabbed an Attwater's prairie chicken, deftly scooping it up before it could scamper away into a corner of the pen. I took notes, and my wife took photos. An assistant took hold of the young bird while Magera performed his health inspection—checking the wings and the feet (healthy feet are essential for a healthy chicken dance). He checked the birds one by one. If I recall correctly, there were about eight individual prairie chickens in the pen we visited that day. It wasn't the only holding pen, but it was the only one we checked.

The prairie chicken health check took about 30 minutes. Then we went back to the visitor center where I conducted a few more interviews and gathered some additional information that I thought would be valuable for my report. There wasn't much to say, unfortunately. The birds' stewards were doing their best to stay positive and optimistic, but the routine setbacks had clearly taken their toll. The manner in which they spoke to me alluded to both frustration and exhaustion.

Attwater's prairie chicken conservation and recovery efforts go back decades. Throughout that entire time, they've gone mostly nowhere.

The birds' numbers boom. Then bust. Then boom again. Then bust again, and on and on in a seemingly endless cycle. This despite the fact that FWS has done everything right. There are no tall structures anywhere at the refuge—birds of prey use them as perches to hunt, and the prairie chickens instinctively avoid tall structures. All the fencing is topped with barbs to keep hawks and other predators from landing, affording the prairie chickens further protection from predation. There's the captive breeding program. And the alternative habitats at the Goliad ranch and Johnson Space Center, which sort of hedge against any potential disaster at the wildlife refuge, like a wildfire. If the population fares poorly in one area, perhaps the birds at one of the other locations would do better and offset any losses, or so the thinking went. But still, the Attwater's prairie chicken numbers are still stubbornly low, far too low to give conservation officials any confidence that the species might be viable on its own. No chance, not today anyway. Without the FWS's caring and nurturing, the Attwater's prairie chicken will go extinct, without question. This is in sharp contrast to the steady if slow upward trajectory of the AWB whooping crane population.

When we visited Attwater Prairie Chicken National Wildlife Refuge for a third and, as it turned out, final time, Magera and his team thought they were finally on the cusp of a real breakthrough. It took some doing, but after a couple years of digging, they said they finally figured out what caused so few chicks to grow to adulthood: fire ants.

This South American invasive species now plagues much of the world. They've even turned up in Japan. In the southern reaches of North America, they are an especially acute problem. My wife complains that she could never enjoy a picnic on the lawn at the University of Georgia, where she went to school, because a fire ant inevitably turned up and bit her. Of course, this insect is everywhere in Texas nowadays. At the time of our visit, FWS had recently determined that a nasty infestation of fire ants at the refuge was holding the Attwater's prairie chickens back. The birds can fly, but they nest on the ground, leaving their young chicks vulnerable. You and I would have difficulty finding a nest given the thick brush and the birds' excellent camouflage. Ants have less trouble spotting them. The fire ants, it was eventually determined, were killing freshly hatched chicks in their nests. So FWS simply added another management tool to their kit: fire ant suppression. By targeting the fire ants, they felt they could protect enough chickens until adulthood, with steady increases in population expected year by year.

Perhaps it's finally working.

As of April 2021, officials with FWS and the Nature Conservancy estimated that the Attwater's prairie chicken population had risen to about 178 individuals surviving at the refuge. "While populations remain at extreme risk, this year's count demonstrates a remarkable turnaround from near extinction in the wild just a few years ago," FWS said in a press release. Still too low for comfort, but far better than just 12 birds. Then again, at approximately 178 individuals, the population of Attwater's prairie chickens at the refuge is about the same as it was back in 1993, per FWS's own admission. Their numbers were possibly much higher at the start of 2023. Hopefully the endless yo-yoing has finally ended, but as with most things in endangered species management and conservation, it's probably too soon to tell. FWS says its ultimate goal is 6,000 breeding adults surviving in the wild.

The uncertainty here lingers on.

Maybe the Attwater's prairie chicken population is finally on the uptick. Maybe it isn't. It's probably just too soon for anyone to tell. And we may never know exactly why this particular conservation effort hasn't seen better population recovery in its entire long history. But there's no harm in asking questions, though they often bring unwelcome scrutiny and may even end up hurting some people's feelings.

If I recall correctly, at the time I had the pleasure to meet with him and tour his refuge, Magera was nearing retirement after serving some 20 years at the Attwater refuge. He and his team kept this species alive and saved it from extinction. For that, they can be rightly proud. But Magera acknowledged where I was going with my questioning and humbly acceded to my point.

After some 20 years of leading this conservation initiative, the Attwater's prairie chicken has escaped extinction and survives, but the effort is stuck in a kind of treadmill—the population numbers just didn't seem to be going anywhere, rising and falling in waves. That's no reason to give up but plenty of reason to rethink things a bit.

We all know the definition of insanity: doing the same thing over and over again and expecting different results. This is what I suspected was happening at Attwater Prairie Chicken National Wildlife Refuge. Through questioning and probing and gently poking, I gathered that the conservation tool kit employed to the supposed benefit of the Attwater's prairie chicken hadn't changed for decades; it was pretty much the same as it was since the wildlife refuge's founding.

There are other common practices that the species' managers employ that I didn't mention earlier. The refuge's managers routinely let ranchers graze their cattle there, the idea being that the cows mimic the ecological function that herds of buffalo once played. There are also prescribed burns, since fire is a natural part of the ecology there, as it is nearly everywhere else in the wild. Invasive plant species are kept under control. But the core formula was the same—breed them at the zoo, acclimate them to the refuge in pens, then release them into the wild and track them from there, hoping for the best.

I asked Magera, as politely as I could, why there hadn't been any changes to the methodology of the Attwater recovery plan. Why wasn't it working? And if wasn't working, why not change things up a bit? I asked him this. Politely and with humility, of course, since I am hardly an expert in these matters.

A few days earlier, I had posed the same question to Mollie Coyn, a Houston Zoo official responsible for caring for the prairie chickens from incubators to the zoo's chicken coops. She acknowledged the premise of the question—for whatever reasons, their efforts just weren't working. But that was no reason to give up, she shot back. "There were a couple of setbacks, but we've got to keep working through them," she told me. "We're committed to them." I agree, of course. But some reflection is, perhaps, also in order here.

Magera's response was more roundabout than Coyn's, but his answer was more or less the same. The frustration was clear in his voice though. One breakthrough, like the new fire ant suppression plan, was met with another eventual setback in the form of drought or too much rain. The birds would get lucky for a few years, and then their luck would run out. At the time, he seemed of the opinion that eventually their efforts would bear fruit. Eventually. If they could only get a lucky break. "We've got the tools in place to grow this population," he insisted.

And he's right.

All the right conservation tools are in place. Designated habitat off-limits to further human encroachment. Captive breeding and wild release. Protection from predators. Regular health checks. Radio-collar monitoring. Population surveys. Assessments. Reports. Funding requests. Oversight from Washington. Public outreach and education. All the tools in FWS's tool kit were on display here, especially habitat management: prescribed burns, grazing, efforts to control invasive plants, animals, and insects, and so forth.

Artificial feeding in the wild?

No, not that tool.

The Attwater Prairie Chicken National Wildlife Refuge has plenty of natural forage for the birds, Magera insisted. In fact, all the officials with whom I spoke about this case study were in lockstep agreement on this point. Bugs, grubs, seeds, you name it. Plenty of chow available to satisfy a wild and hungry prairie chicken, especially considering that there are so few of them—they need not worry about competition for forage with other Attwater's prairie chickens. Feeding birds living wild and free in an open expanse like Attwater Prairie Chicken National Wildlife Refuge made no sense, I was told. Yes, they're necessarily fed in captivity, the experts all acknowledged, but not out in the wild. There, the birds had all the sustenance that they could ask for. And at any rate, they needed to learn to fend for themselves. Luckily, this isn't hard. Coyn said the prairie chickens don't seem to get acclimated to or too dependent on the humans, even during their time of captivity. There were no signs of creeping domestication.

But maybe, just maybe, the circumstances have changed. Maybe the unexpected should be expected. Perhaps feeding them in tough times, like droughts, would elicit a more positive and sustained population response, as was seen with the red-crowned crane.

Perhaps the habitat that Fish & Wildlife is struggling to protect for this particular critically endangered species is radically different from the coastal prairies that were once home to more than one million Attwater's prairie chickens 300 years ago. At one point in our conversation, Magera mused that the fire ants could be chasing out populations of other insects that the Attwater's prairie chickens might otherwise feed on, especially their chicks. And maybe there are other changes that have gone undetected.

By all means, get rid of the fire ants. But is that enough? And how can we be sure that eradicating the invasive fire ants won't mess something else up? Perhaps they're now serving some kind of niche function or ecosystem service that we're not yet aware of (though I doubt this).

Let's put all that aside for now. Despite my somewhat harsh assessment, I think it's fair to count the Attwater's prairie chicken conservation program as one of ESA's success stories, albeit a modest one. After all, this species is not extinct. It's hanging by a thread, for sure, but this little dancing prairie chicken will be with us for some time to come, surviving another day to frustrate future visitors to the refuge hoping to catch at least a glimpse of one before ultimately turning home disappointed.

⁓

"It's beautiful, isn't it?" Jonathan Jarvis asked me as I fumbled with my voice recorder and notepad.

"It is," I agreed, struggling as I was to look composed and put together. At the time, I was considering how best to manage a standing interview outdoors, recording audio while jotting down notes. I would need a third hand. Ultimately, I put the notebook in my jacket pocket and focused solely on the interview and recording clearly. We were high up a mountainside, so I had to make sure the wind didn't muffle our voices or otherwise make it difficult for me to hear what was said. I could always transcribe the entire thing anyway, I reasoned. I wasn't on any firm or tight deadline.

The beauty Jarvis was referring to that day was a breathtaking vista that engulfed us, the treeless expanse of tall mountain peaks seen from the summit of Trail Ridge Road in Rocky Mountain National Park, Colorado. The view was and is stunning, though the mountain peaks weren't visible around us in 360 degrees, not from where we stood. The view stretched more like 300 degrees around us. To get the full spectrum of the southern Rockies' magnificence, we would have had to walk to the tallest point of the mountain looming next to the parking lot where our tour van was parked. The air was chilly and thin, and I was already out of breath, the consequence of living so much of my adult life at sea level. And I knew the wind would get worse the higher we went, further complicating my efforts to get a clear recording. My wife was higher up the mountainside already, happily snapping photos, but the wind wouldn't be a problem for her. It was cold. My jacket proved insufficient for the conditions at the top of Trail Ridge Road, though it was just fine back in Estes Park or at lower elevations. Jarvis noticed my shivering and kindly suggested we talk at a slightly lower elevation behind some boulders to get out of the wind. I thankfully agreed.

I've met two Secretaries of the Interior. Sally Jewell led the Department of the Interior under President Obama. I spoke with her at an energy conference in Houston and then again by phone when she called from her office in Washington. At most, I think I had 15 to 20 minutes with her in one sitting. Later, I would meet and speak with Ryan Zinke, Interior chief under

President Trump. Zinke and I spoke at that same energy conference, which is more of an annual trade show devoted to offshore oil and gas technology. He gave me about 15 minutes of his time, maybe a bit longer. Both DOI chiefs were affable enough, but they were both all business. Their speech was guarded and calculated, mindful as they were that they were speaking to a journalist and on the record. When they spoke, those two kept it professional and rarely deviated from the script. They were constantly on message.

Jonathan Jarvis is the only director of the National Park Service that I've ever had the pleasure of meeting and speaking with. We've conversed on two separate occasions and for far longer periods of time than that spent with Jewell and Zinke. I found both opportunities to speak with Jarvis to be entertaining and informative at the same time. Our discussions were more casual, if somewhat wonky, conversations, like back-and-forth banter between two newly acquainted colleagues who found instantaneously that they could get along and work well with one another.

Chances are good you know Jarvis from the grilling he took from members of Congress back in 2013. That year, the Republicans and the Democrats were at each other's throats as usual and for the usual reasons. Their dispute eventually culminated with a shutdown of the federal government, since the two sides couldn't agree on spending priorities and authorizations. For a time, all federal offices were forced to close, since there was no way to pay to keep them open. The National Park Service is a federal office. It, too, was forced to close, as were all the national parks in the country, just as millions of Americans were hoping to visit them. You can imagine how well that went over.

Chagrined, angry American park afficionados by the cruise-ship load called their representative congressmen and congresswomen to yell at them. These flustered congressfolk needed someone to yell at in turn, so they picked Jarvis. He was called in before the lawmakers in Washington so they could be seen on television yelling at him and blaming him for the park shutdowns. Members of Congress quite literally hauled Jon Jarvis before them so they could harangue him for doing what they had ordered him to do, all on live television. This is probably where you've seen him before, if at all. He took it all in stride, if I recall correctly. Like I've said before, Washington is an interesting town.

We met a year prior to all this happening.

Jarvis cuts an entirely different figure from Jewell and Zinke. As I noted, his style seemed more casual and conversational. On both occasions when we met, our discussions went on longer than either of us realized or planned. The second time we met, Jarvis had flown in from Washington to Houston to inaugurate a new urban canoe trail. That initiative was part of the National Park Service's efforts to bring the national park experience closer to cities, both to encourage healthier outdoor activities and, it was believed, to take pressure off the increasingly crowded and strained park system. I spent a couple of hours with him then. He later invited me to go on a canoe trip with him and a group of volunteers down a bayou in north Houston—Greens Bayou, to be precise—but I was busy with other things that afternoon.

My first meeting with Jarvis took place at Rocky Mountain National Park, in the high peaks and chilly mountain air during a reporting trip. I spent the entire morning with him that day. He was decked out in his park ranger uniform as usual, wide-brimmed Smokey the Bear hat and all. Back then, he sported a big bushy moustache. He probably still does. The look in his eyes always suggested he was very much a glass-half-full type of person—upbeat, affable, and positive. He smiled wide practically the entire time we spoke, a big toothy grin, nodding approvingly at the Park Service staff and volunteers busy surveying and assessing the sparse tundra vegetation at their feet, the only kinds of plants that can survive at that altitude and in those conditions.

Jarvis and I spoke about climate change. And about BioBlitz, the event that brought me to Rocky Mountain National Park that day in the first place. BioBlitz is the Park Service's "citizen science" initiative. *Citizen science* refers to the idea of eliciting regular folks in the service of scientific research. It's proven to work, if designed and done right. At the top of Trail Ridge Road, officials had gathered to direct volunteers in the collection of data on high-altitude plants and plant communities, part of a longer-term effort to monitor the impacts of climate change. They were also hoping to spot signs of the elusive pika, an endangered species and relative of the rabbit, which inhabits the highest points of mountain ranges (a species of pika can be found in Japan, as well, and it's just as rare as its North American cousin). Later, I would accompany Jarvis to a lower and far more forested part of the park to watch officials direct elementary school students on a separate citizen-science data-collection effort.

We also talked about the pika.

This little furry animal is most closely related to rabbits, except that it's found only at high altitudes, especially mountain peaks above the tree line, whereas rabbits are found pretty much everywhere in lower-lying climates. Pikas are found in the mountain peaks of Hokkaido, as well. They are seen as among the species most at risk to climate change: as temperatures rise, ecosystems experience climatic shifts, and those highland peaks are starting to become a little warmer for longer periods of time. Pikas thrive in colder environments, so conservationists are concerned that the changing conditions will make life inhospitable for the pika, and there's no place to escape, given that they already inhabit some of the highest places on Earth. Gradually warming conditions also could bring more humans to pika habitat, threatening their survival that way, as well.

Jarvis expressed particular concern for the pika and its long-term survival. That day off Trail Ridge Road, however, what he really wanted to talk about was the new direction that the National Park Service was setting out on for its 100th anniversary. Decades of outdated policy were being overturned; a philosophy first codified and adopted in 1963 was starting to get left behind. A new ecosystem management philosophy was taking hold.

National parks, like Rocky Mountain, are increasingly coming under stress. Climate change and climbing numbers of visitors inflict fundamental and permanent changes on parks everywhere. When I was growing up in Colorado, Rocky Mountain National Park was overwhelmingly populated with coniferous trees. It was mainly an evergreen environment. Then global warming came along and helped to spark a nasty infestation of the mountain pine beetle, a parasite that kills pine trees. By the time of my visit, the Rockies had indeed become a purple mountain majesty—the hills and mountainsides were rendered purple from all the dead trees. The patches of green between them were deciduous species taking advantage of the gap in the ecosystem created by the pine beetle. Rocky Mountain National Park is transforming into a mixed forest system, and the Park Service ultimately decided to just let it happen. They couldn't stop climate change, after all. The best they could do is try to keep up with and manage the transformation already underway. When I was there, this mainly meant felling dead trees closest to roadways to keep motorists safe.

Americans began invading the national parks in earnest following World War II and during the nation's postwar economic boom. In the 1950s, the lines of separation between humans and nature at the parks were basically blurred. Park managers took a casual view of how visitors comported themselves. Taking souvenirs was common—rocks, pine cones, sometimes whole plants. Visitors got close to the wildlife, oftentimes a little too close. They even fed the bears. This, of course, led to some unpleasant conflicts between humans and nature. Bears and other wildlife began associating the human visitors with food. You can imagine what that led to.

By the early 1960s, the Park Service was undergoing some serious soul-searching. Then in 1963, a conservation official by the name of Aldo Starker Leopold drafted a now-famous report titled *Wildlife Management in the National Parks*. This one document set the tone for how the National Park Service would approach human–wildlife interactions for decades following.

Leopold basically argued that a sharp dividing line should separate the parks' nature from the parks' visitors. Top officials agreed, and more or less implemented all of Leopold's recommendations.

Taking souvenirs was banned. Visitors were strictly kept to designated trails and scenic overviews. Backcountry activities had to be approved and permitted in advance. And most important of all, feeding the parks' wildlife was strictly banned. As Jarvis described it to me, Leopold argued that the Park Service should strive to keep conditions at parks as close as possible to a "historical vignette" of what they most likely looked like before the humans came. "A snapshot of history" is another phrase Jarvis used to describe the thinking to me. Leopold argued that it was the Park Service's duty to maintain the national parks and the wildlife that live in them in a state as close as possible to conditions as they existed before European settlers arrived to North America.

It worked, for a while at least. Today? "This is an illusion," Jarvis told me flatly.

He pointed to the lichens and tiny shrubs at our feet, the high alpine vegetation found only here and in the Arctic tundra. With global warming creeping in, it is no longer possible to maintain this sensitive ecosystem as a snapshot of history. New thinking was required here, he said. A new philosophy has to take hold. This vegetation would undergo changes as the climate changed, and the best the park could do is monitor these changes and respond to them as necessary and as best they could.

In 2012, the National Park Service's advisory board and scientific committee issued a new report titled *Revisiting Leopold: Resource Stewardship in the National Parks*. This new document declared that it was time for a radical change in approach. In other words, it was time to rethink things, the report concluded. "New knowledge and emerging conditions—including accelerating environmental change, a growing and more diverse population of Americans, and extraordinary advances in science—make it urgent to reexamine and, if necessary, revise the general principles of resource management and stewardship in the national parks as described in the Leopold report."

The National Park Service would no longer endeavor to maintain these ecosystems as historical vignettes or as snapshots of history. Rather, it's now been directed "to steward NPS resources for continuous change that is not yet fully understood," per the newer report. In other words, park managers must expect the unexpected and accept that the old ways of doing things may no longer work. They're encouraged to experiment. And they've been directed to involve the public in bigger and more meaningful ways. We still can't feed the bears—or any other wildlife for that matter—and this is entirely sensible. But the hard dividing lines between humans and nature are now starting to become a bit blurred. NPS now accepts that change is inevitable and that their approaches toward natural resource management must be flexible and adaptable in the face of these inevitable changes.

Jarvis laid out this new NPS mindset to me cheerfully and optimistically. He nodded toward the road below us. It was a weekday and still early in the season, but already Trail Ridge Road was becoming noticeably congested with family sedans and RVs. The parking lot appeared to be filling up fast. The east entrance to Rocky Mountain National Park is only about a 90-minute drive from downtown Denver, then one of America's fastest-growing metropolitan areas. Denver's air pollution easily wafts over the park, and its chemical traces find their way in the waterways there. Things have changed considerably from what they were in 1963, Jarvis told me. The Park Service is trying to foster national-park-like experiences in the cities because the cities are fast creeping up on the national parks—it's a desperate bid to beat back a rising tide. A lot of the changes to the parks we haven't noticed yet. Some of them, we have. And more changes are in store. So it's important to be flexible and adaptable, he reckoned. NPS would no longer be rigid in its thinking and park management approaches, Jon Jarvis assured me on that mountain that day.

The changes aren't radical.

Old restrictions on visitor activities are still in place. NPS is still guided primarily by the precautionary principle, a conservative approach to conservation that relies heavily on the best available scientific evidence. But where warranted, different approaches or management strategies should be allowed and even encouraged. As its 2012 report declared, NPS "should formally embrace the need to manage for change" and in an ongoing, never-ending process of reevaluation. "This will require concerted examination by NPS professionals and stakeholders, as well as the relevant scientific, legal, and policy analyses," the report states.

The National Park Service's 100th anniversary prompted this soul-searching and new thinking that Jarvis described to me. It didn't have to occur simply because a milestone year had been reached. In fact, a broader policy reflection and reconsideration initiative had been underway for several years prior. But ultimately, NPS looked back on its 100-year history and the transformational Leopold report of 50 years prior and voted to break with its past, forging a new path ahead. The national park system's 100th birthday was simply as good a time as any to announce this new thinking and new direction.

And here we are at another milestone.

Fifty years ago, erstwhile political adversaries became allies and passed, in a bipartisan and cooperative fashion, the Endangered Species Act. It's working fabulously in some places, not so fabulously in others. And compared to 1973, the threats to biodiversity existing today are more varied and more severe than ever. The current politics in Washington also spell trouble for the ESA, I'm afraid. The federal government's debt is massive. Pressures on spending and financial resources allocations will mount. The best lobbyists will win the day, and I fear the wildlife can't necessarily afford the best representation on Capitol Hill. Some officials openly muse that some species should be allowed to go extinct simply because conservation resources are limited and the chances for these species' recovery is low. This "let 'em die" attitude ultimately may succeed because the American public's attention is, at present, diverted elsewhere, as is the press corps'.

The world keeps changing in unpredictable ways. As I write this, gasoline prices are at record highs. Inflation is biting into household budgets badly. Rising population and urbanization in the United States have delivered a sharply rising cost of living, and Americans' wages aren't keeping up. The nation is badly divided politically and culturally, and this rift is widening and worsening. During all this, an international team of wildlife champions have been trying like mad to stir public interest in their efforts to strengthen the UN Convention on Biological Diversity and to reinvigorate the global battle against biodiversity loss, a battle largely being lost. The science communicators are failing.

The television news has covered none of their initiatives or press releases, as far as I can tell. The broader public shows scant interest. A saga over Elon Musk's purchase of Twitter is garnering far more attention. The Intergovernmental Science-Policy Platform on Biodiversity and Ecosystem Services, or IPBES, has just issued a report calling for changes in the way communities everywhere manage and exploit wild species. This initiative isn't getting much press or public notice, either. The press inattention I described earlier in this book is still very much driving the dynamics. Clickbait and rage porn have conquered the news cycle, and the steady death of America's biodiversity (and of the world's) just isn't getting the attention that it deserves.

But the monied lobbyists, the clickbait, and the divisions riling American society may not completely dominate the way contemporary endangered species policy is legislated and enacted on Capitol Hill.

In June 2022, the U.S. House of Representatives passed, in largely bipartisan fashion, the new Recovering America's Wildlife Act. The new act directs $1.3 billion in annual designated funding to the 50 states' wildlife management plans in a bid to halt and reverse a noticeable decline in biodiversity, especially at freshwater habitats. The Wildlife Society says the new law prioritizes Species of Greatest Conservation Need, or endangered species known to be in decline in the wild. The bill is pending in the Senate as of this writing. Conservation nonprofits unanimously approve of the bill and seem hopeful that it will eventually become law. It should pass. The latest news reports say there are enough Republican cosponsors in the Senate to see it through. Or maybe it won't—the Senate has disappointed us before, after all. It doesn't take much to distract senators, and many of them are almost certainly on the lookout for useful distractions. And the American press corps is certainly out to please them in its never-ending quest to "give the audience what it wants" rather than what it needs.

Back in 2012, those responsible for managing America's national parks and national monuments decided that a change was in order. Maybe the stewards of the Endangered Species Act need a new approach, as well.

When it comes to endangered species management, the old approaches won't be abandoned or overturned completely, nor should they be. But this milestone moment for the ESA will no doubt lead to some new official reviews and some new soul-searching. This is a very good thing. Some flexibility and new approaches may be in order. Including—perhaps (and by no means definitively)—sustained artificial feeding of the sort practiced in eastern Hokkaido, Japan, in some circumstances. But not in all circumstances, of course.

Or not.

I'll defer to the experts here.

Diplomats are again busy trying to beef up the Convention on Biological Diversity (CBD). At the end of 2022, CBD negotiators concluded something called the Kunming-Montreal Global Biodiversity Framework agreement. Governments have agreed to an aspirational goal of setting aside or otherwise implementing species habitat protections on at least 30 percent of the world's land and waters by 2030. They've also promised to spend a lot more money on species protections. All positive signs, but I'm not holding my breath. After all, as Deakin University professor Euan Ritchie said in a reaction statement issued by the Australian Science Media Centre, "Words and intentions are one thing; it's corresponding and genuine action and meaningful, measurable outcomes that matter."[8]

The previous large-scale attempt to strengthen the CBD didn't turn out all that great, but delegates seem determined to make a difference this time around. Many of the sticking points are the same, however. Nations are at odds on how to collect and share genetic information—in other words, the most equitable ways for exploiting biodiversity, rather than saving it. Rich nations want to own all the rights to patented products developed from the genetic information collected from wild plants and animals. Poorer nations think these riches should be shared, especially when that genetic information is obtained from species native to their lands or from species found in areas of the open ocean under no single nation's jurisdiction. Governments may ultimately set this question aside for a separate track of diplomacy while they work on a final outcome document that they can all live with.

The Convention on Biological Diversity doesn't ask much of its signatories; that's why there are so many of them. There are no firm conservation targets. No quotas on natural resources extraction are imposed. No fines are levied whatsoever. The CBD's function is to facilitate information sharing and to encourage states to protect wild plants and animals, particularly endangered species. It can hardly be any other way because, as I explained earlier, to increase the odds of drafting and successfully passing a multinational environmental agreement, that agreement and the framework it builds must meet three conditions: participation must be inexpensive or free, participation must require little from participants, and there must be absolutely no penalties for noncompliance. But as with so much else at the UN, the CBD finds itself mired in other arguments, namely those concerning the potential for future profits to be had from exploiting wild plants and animals.

Most governments want the world's biodiversity and the digitally sequenced genetic information acquired from it to be considered the world's common heritage and to be used in a way that benefits them all (the governments, of course, not you and me). Richer governments and the corporate interests that own them very much disagree. So even something as seemingly uncontroversial as "hey, everyone, let's stop driving thousands of animals and plants to extinction" becomes mired in the messy business of global politics. "The 4th negotiating session in Nairobi in June was intended to be the last one but an agreement is not yet there," the CBD secretariat's executive secretary Elizabeth Maruma Mrema said in a press release announcing an additional Montreal negotiations round. "The parties will have to decide in December whether digital sequencing information will be part of the framework or dealt with separately." They decided to kick the can down the road when it comes to this dispute.

Back in the United States, the Fish & Wildlife Service is considering extending Endangered Species Act protections for several species not yet listed, even as there's talk of upgrading the status of or even delisting the whooping crane (again, it's far too early for this in my opinion). According to the Wildlife Society (TWS), a court is ordering FWS to consider placing wolverines on the endangered species list. A separate court ruling said FWS should consider listing sage grouse populations in California and Nevada as threatened. FWS declined to do anything about wolverines and the sage grouse back in 2020.

FWS is also seriously considering a listing for the plains bison inhabiting Yellowstone National Park. Apparently, new residents are flooding into communities surrounding the park and development is creeping in. FWS worries that hemming in the bison's range will inhibit the species' recovery. Conservationists are increasingly worried about Florida's manatees. Several apparently starved to death after water pollution from the booming state (now experiencing rapid population growth as newcomers rush in) killed off seagrass beds, food critical to the manatees' survival. To their credit, state and federal conservation authorities stepped in with emergency feeding, according TWS, tossing some 200,000 pounds of lettuce at the manatees to get them through the 2021–2022 winter.

Speaking of bison, the European variety is doing better as of late, according to the International Union for Conservation of Nature (IUCN), which upgraded the European bison's Red List status from "vulnerable" to "near threatened" a couple years ago. That species was once extinct in the wild and survived only thanks to captive breeding. But this is hardly a success story. At slightly more than 6,000 surviving individuals, IUCN says the European bison today exists in 47 "free-ranging" herds in Eastern Europe, with all roaming herds isolated from one another. IUCN says most of these herds are not genetically viable long term without continual human intervention. "The species remains dependent on ongoing conservation measures such as translocations of bison to more optimal open habitats and reduction of human-bison conflicts," IUCN says.[9]

Monarch butterflies, or the migratory subspecies, is faring worse. IUCN declared North America's monarch butterfly an endangered species in a July 2022 media release. Development and habitat loss are the main culprits, but IUCN also blames climate change. Humans have reduced monarch butterfly numbers by up to 72 percent by some estimates, and the population is on a downward trend. They're losing their wintering habitat, and American farmers are killing them with pesticides. Millions are probably killed each year by cars as they cross highways splicing through fields of flowers. As America's population has swelled, especially in the West, monarch butterfly populations have plummeted. If we want more wild European bison and migratory monarchs, we may need a lot fewer Europeans and Americans. I have more to say on that subject much later.

Meanwhile, a team of scientists is arguing that invasive reptiles and amphibians have damaged the world's ecosystems to the quantifiable tune of $17 billion since 1986. The American bullfrog and brown tree snake are particularly damaging, they said. Their study, published in July 2022 in the journal *Scientific Reports*, arrived at that monetary estimate largely by looking at crop damage caused by these invasive species, but the global spread of non-native species is wreaking havoc on native flora and fauna, as well, likely pushing many to the brink of extinction. The cost of all this devastation can be assessed in dollars and cents. Scientists often put ecological damages in monetary terms in an effort to pique the interest of governments and the press. Most of the time, it doesn't work, but you can't fault them for trying. "A greater effort in studying the costs of invasive herpetofauna is necessary for a more complete understanding of invasion impacts of these species. We emphasize the need for greater control and prevention policies concerning the spread of current and future invasive herpetofauna," the scientists argued in their report.[10]

A separate study published in that same journal around the same time shows the world is doing a very poor job of ensuring the survival of endangered carnivores. Of 362 large endangered carnivore species assessed, scientists say only 12 of the cases showed noticeable improvement in conservation status. The rest—350 endangered carnivores covered in this one study—all still teeter on the brink of extinction, despite all these species being subject to some form of conservation regime or statute, whether it's the weak CBD or firmer domestic legislation. Only 3.3 percent "have experienced genuine improvement in extinction risk, mostly limited to recoveries among marine mammals," those researchers wrote. The solution, they say, is even stronger laws and greater resolve from governments to actually enforce the laws. Despite existing legal protections, far too many iconic creatures face extinction. The report describes the decline of South Asian vultures as "catastrophic" and in need of urgent attention. Still, this particular study found positive correlations between the presence of conservation laws or a conservation regime and endangered species recovery. The legal regime matters, as do the conservation practices codified, permitted, or mandated under that particular legal regime. "In the face of an accelerating extinction crisis, *scientists must draw insights from successful conservation interventions to uncover promising strategies for reversing broader declines*," that study argues.[11] My emphasis.

I couldn't agree more. Let's take another look at one such conservation intervention, one of the most successful ever undertaken during this ongoing worldwide extinction crisis.

Time to check in with *tancho* again.

CHAPTER EIGHT

"Could" versus "Should"

"Are you sure we're allowed to be back here?" my wife asked.

"Of course," I answered, most confidently. "It's on the map, isn't it?"

The trail was on the map. And it was clearly a footpath. Sure, it wasn't paved, but not every nature walk in Japan is cleanly paved, or marked, or laid with wood planks, or otherwise developed. Sometimes, they're just well-trodden footpaths, as this was.

"Yeah, but this seems a bit away from the main area," she pointed out.

She wasn't wrong. But we weren't that far into the bush. East Hokkaido can be wild in places, but it wasn't exactly the Australian Outback. It was old farmland long returned to nature, to the ecosystem that surrounded the creeks and river we were ostensibly headed to, in the backyard of one of Hokkaido's most famous tourist hot spots.

The Akan International Crane Center is where intervention feeding for Japan's population of red-crowned cranes first began. The annual winter feeding regimen continues there to this day, although the site itself has been greatly improved and given a more formal role. Previously, this was just an empty field to toss corn around, an abandoned farm surrounded by hills resting silently in a vista guarded closely by the looming peaks of two ancient volcanoes to the north, Meakan and Oakan. The locals liken this volcano pair to an old married couple. Oakan is the old man, stubborn and set in his ways, quiet and never changing season after season. His wife, Meakan, is the livelier and very much still active volcano, the hothead of the pair. During my time there, the Akans—the old married couple volcanos—could be spotted in the distance nearly every day. Look closer, and you could see steam pouring out of Meakan's peak. Oakan must've said something stupid to make her mad again, I'd think.

For most of its history, the Akan center was just an empty fenced field. Later, the authorities added a building to serve simultaneously as a research base, miniature museum, reference library, and bookstore. The center itself isn't that old—it was officially established in 1996, just one year after I first visited Japan. The people in charge probably figured that it would be better for everyone if they started controlling access and even charging for admission, creating a steady stream of revenue they could tap into to pay for the care of the red-crowned cranes while generating a few jobs in a region badly in need of employment opportunities. Today, the Akan International Crane Center is a well-known East Hokkaido destination. A rest area now lies just across the street where visitors can drop by to get a quick bite to eat or savor the region's famous to-die-for ice cream made with local dairy (highly recommended if you find yourself in the area).

The center sits at the north end of the village of Akan, which used to be its own municipality but has since merged with the city of Kushiro. My wife and I seriously considered buying a house in Akan before we found our current place. Akan is slightly warmer than Kushiro in winter but hotter in summer. I'm not sure why. It's a bit snowier there in winter, too, but this portion of southeast Hokkaido receives the least amount of snowfall on the island, and Akan lies within this snow shadow. It gets noticeably snowier and colder at higher elevations, including at Tsurui. Admission to the center is cheap, cheaper still if you come in a larger group, and open year-round.

The Akan Center's history is similar to that of the more famous feeding station in Tsurui. As the center's website tells it, one day a man by the name of Yamazaki Sadaijiro grew worried about the future of the red-crowned cranes and decided to take matters into his own hand. He noticed cranes resting at his farm, and they looked hungry. So one day in January 1950, Yamazaki famously walked to a portion of his fields and scattered some leftover corn kernels to the handful of cranes that he found there that day. It was the same feed he used to fatten up his dairy cattle—near as I can tell, this is the only reason why the seasonal feeding centers on corn.

At first, the cranes were hesitant. Having suffered decades of abuse at the hands of humans, *tancho* wasn't about to simply throw caution to the wind. Yamazaki had to keep his distance before they would even touch the corn. But eventually the cranes gobbled it up and asked for more—from a safe distance, of course. Yamazaki decided he needed to build their trust and make this feeding a new routine. So every morning at a set time, he showed up with his bucket of corn/cattle feed, scattered it on the ground, then walked away. Eventually, the cranes got used to his presence and this morning routine. The experiment worked—the red-crowned cranes can be fed, Yamazaki discovered. As the center's website retells it, "Artificial feeding was a success." From this point on, the center argues, rates of starvation among red-crowned cranes declined sharply despite the region's harsh winters, and their population numbers began rising steadily.

In winter, visitors are restricted to the building and to an area behind fencing meant to keep the humans out so the cranes can spread their wings, rest, and fill their bellies undisturbed. The Akan Center easily hosts more than 100 photographers any given winter day lined up behind the fences, waiting patiently to snap pictures of the iconic red-crowned crane dancing in the snow. In the warmer months, however, the feeding stops and the cranes head back to the food and shelter of Kushiro Marsh and other wetlands in the area. When it's safe to do so, the center opens its vast backyard to casual day hikers, like my wife and me, looking to enjoy the weather and get a better sense of what the center's grounds look like up close.

Trails lead visitors around the perimeter of the feeding grounds. Farms lie to the north and south, but visitors are treated to expansive views of verdant hills all around them. The Akans loom, as well, though they may be difficult to spot from this quasi-valley. Tall grasses give way to forest to the east, signaling the approach to the Akan River. Get close to the foot of the forest to see interpretive signs telling you the kind of trees you're looking at. Japanese alder is found here, but that's fine since this isn't technically in the critical wetland habitat. Still, this part of the center is the least hospitable to the red-crowned cranes because there are so many places for predators to hide and too much brush to let the cranes beat a hasty retreat if necessary. My wife told a tale about when she visited the center in the wintertime by herself. A wayward crane had inadvertently stumbled into the forest, perhaps a younger member of the flock. The crane's buddies waited outside. She said the other cranes seemed anxious and worried for their friend. Eventually, the lost crane found its way back to the group and to the open area without incident, and they all relaxed and continued pecking at the corn in the snow.

If you follow the trails away from the visitor center, you eventually come to a fork in the road at the foot of the forest. Two paved paths continue toward the main area, but a third unpaved trail leads into the trees. My wife thought the latter was off-limits, but I disagreed, pointing to the obvious trail, the fact that it's on the map, and to an explanatory sign resting before a tree at the start of this fork. We walked a little way in and found a pleasant creek. We suspected that this was the area where my wife saw that lost and wandering crane the prior winter—the creek must've been the draw.

I was having a blast working my way through the forest, hoping we'd make it to the Akan River and as far away from the other guests as possible. The trail continued but narrowed. We made our way across the creek and a bit further east then stopped in our tracks. "*Anooo*," we heard a woman say loudly, clearing her throat just behind us. "*Sumimasen.*"

"*Hai*," my wife responded. I turned around.

There stood Kawase Miyuki in her tall green rubber boots and park uniform, a worried look on her face. I was a bit annoyed—yes, we know there are bears in the area, I was about to explain to her, but we're making plenty of noise and really don't expect to encounter any that day, as there were no reports of any recent sightings. When folks decide to go hiking anywhere in Hokkaido, they're generally careful to check the bear reports, just as they would the weather reports, and my wife and I had taken up this habit, as well. The rest of Japan has black bears, aggressive but small enough that a guy my size could probably scare one away. Not in Hokkaido, home of *higuma*, the 600-pound brown bear monster cousin of the Alaska grizzly bears.

"*Anoo*," Kawase continued in Japanese, "you're not allowed to be back here."

"I told you so!" my wife snapped at me.

"Not supposed to be where?" I asked.

"Anywhere back in this area," Kawase answered politely but sternly. You're supposed to stay on the trails."

I protested. "We *are* on the trail. And this is on the map."

"It is?" Kawase seemed shocked.

"Yes." We showed her, adding that a larger map on a placard at the trailhead showed the same trail leading into the forest (though, admittedly, none of the maps seemed to suggest that you can go all the way back to the river).

"Oh, I'm sorry," she apologized. "But that's a mistake. You're only allowed back here with an escort. We're worried someone might get injured."

This wasn't exactly bushwhacking through the Amazon, but the center does receive a lot of elderly visitors, so I suppose someone on high decided to make a new rule reducing the fun quotient for visitors like us to help alleviate any liability the park might face should someone's grandmother slip and break a hip. I'm already used to the way America's lawyers have lawyered the fun and spontaneity out of everything. But Japan, too? No big deal, I suppose.

We bowed and apologized and dutifully followed Kawase back to the approved zone. Two other guests stood there waiting for her—a couple about our age apparently on a guided tour. "I'll talk to you later, OK?" Kawase said to me.

"Yes, I'm looking forward to it," I told her. And I was.

Kawase Miyuki is manager and lead ranger at the Akan International Crane Center, or at least she was at the time of our visit. I'm sure she's still there. Kawase is a naturalist and expert on the red-crowned crane, and she leads the winter feeding regimen. She's also a local celebrity. She's appeared on television episodes featuring the attractions of East Hokkaido, and the news media have interviewed her on numerous occasions to discuss the status of the area's red-crowned cranes and their ongoing recovery. I had already made an appointment to speak with her, which would be our second occasion to converse. We first met when I interviewed her in winter 2019. I was looking to do a follow-up that day we were heading to the Akan River, but we arrived early to check out the trail system before getting Kawase's update on how the cranes were faring now that the Ministry of the Environment was enforcing a new experiment—reduced annual artificial feeding volumes.

We walked the grounds a bit more, taking photos and remarking on the little details we discovered as this once farmland steadily returned to nature. About 30 minutes later, we made our way back to the visitor center for a quick chat with Kawase. I didn't have many questions. I was just curious about how things had been going since we last spoke and whether the government was beginning to rethink its plans for *tancho*.

A year before I relocated from Texas, the Red-Crowned Crane Conservatory (RCCC) estimated that about 1,500 *tanchos* populated the wetlands and open fields of East Hokkaido, with the largest concentration keeping to Kushiro Marsh and its immediate vicinity. By the winter when I first met Kawase, that population estimate had risen to more than 1,800 birds. During my first year of research, the winter count hit 1,900 cranes. However, the new methodology employed by RCCC to reach these population estimates reveals that the authorities don't actually have a firm handle on how many red-crowned cranes now live in Hokkaido.

For reasons that escape me, beginning in 2013 the folks responsible for the annual red-crowned crane population survey decided that they would start publicizing their estimated number of surviving cranes by rounding up or down to the nearest 50 birds. Say RCCC authorities and their helpers counted 1,726 cranes during the survey period. They would report this in their statistics as 1,750 birds. Conversely, if they confirmed only 1,724 cranes that season, then this figure got rounded down and the official stats would be recorded as 1,700 red-crowned cranes.

This seems a rather foolish way to go about things, considering that the entire estimated population was barely 30 cranes when Mr. Yamazaki started tossing his corn at them so many years ago. But the Japanese conservationists seem to prefer this ambiguity, especially since population numbers of any given species don't remain static, and even the best survey efforts fail to count every living member of a population. Some well-informed corners think China's human population is closer to 1.2 billion rather than Beijing's official government estimate of 1.4 billion, suggesting an overcount of almost 200 million people. Now that's a bad rounding error. It's even harder to count wild animals since they don't respond to census surveys. A bear expert at Hokkaido University once confessed to me that he and his colleagues have no firm idea how many *higuma* actually reside on the island, with the exception of the Shiretoko Peninsula, home to Asia's largest concentration of brown bears.

Kawase told me that, by her guess, the red-crowned crane population in Hokkaido is somewhere between 1,800 and 2,000 birds. That figure probably will be closer to 2,000 by the time you're reading these words, with any luck, though that's just my guess. What we do know for certain is that after decades of stagnation and struggle, the red-crowned crane population began steadily rising at healthy annual rates beginning with Yamazaki's historic intervention. Where the whooping crane struggled, the red-crowned crane soared, but only on Hokkaido—a separate migratory population found in China, Siberia, and parts of Korea has been in steady decline and is now estimated to be below the population numbers seen in Hokkaido, remarkable given that the Asian mainland population has a lot more habitat at its disposal.

After Akan, two more feeding stations were established by the locals: the Itoh sanctuary and Tsurumidai, which is an open field off the highway at the southern edge of Tsurui and across the street from Do-Re-Mi-Fa-Sora, a popular café (the name Tsurumidai roughly translates to "crane viewing platform").

During my first interview with Kawase, we sat down together at the table in the middle of the visitor center's main hall, adjacent to the check-in counter and ticket booth. I saw several books around me that I had in mind to purchase but later forgot. She had just appeared on a television special broadcast earlier in the week, and I could tell she was well used to dealing with the media. Joining us was Matsumoto Fumio, a biologist and crane expert with the Kushiro Zoo. Matsumoto happened to be at the center at the time of my visit and had some time to spare, so he kindly agreed to sit in on my discussion with Kawase and provide some answers of his own.

I explained to them both my investigation into the history of the Aransas-Wood Buffalo whooping crane population and the struggles of wildlife managers there trying to increase whooper numbers enough so that the species no longer requires active intervention by well-meaning humans. I explained to them the setbacks caused by droughts. Most of all, I wanted to explain to them how closely the two conservation initiatives parallel one another, with the exception being that the AWB whooping crane population had barely reached 500 individual birds by the time I landed in Japan. Meanwhile, by Kawase's expert estimate, the red-crowned crane population was triple or almost quadruple the size of the flock wintering at Aransas. The two programs were comparable back in 1950, but since then, the Japanese have lapped the American effort.

I explained to them all that I've told you earlier in these pages. Both species are similar to nearly indistinguishable in terms of morphology, behavior, biotic potential (the potential for the population to grow through breeding), and longevity. They are different in plumage, of course, but similar enough that my students have a hard time telling them apart, despite the fact that red-crowned cranes feature prominently in Japanese art and popular culture. Like I've already said, I'm willing to bet these two species are closely related enough that they could interbreed. They're certainly more closely related than a horse and a donkey. But this is an experiment that, thankfully, no one will allow me to conduct.

The AWB whooping crane undertakes a grueling and long migration throughout central North America twice a year. But the red-crowned crane must endure bitterly cold winters every year, stubbornly refusing to migrate to warmer climes further south. Whooping cranes face some threat from predators, but so do red-crowned cranes. The whooping crane also has a vastly larger area of federally protected lands available to it. Red-crowned cranes have far less land to stretch their wings, and what they do have has been shrinking due to the past drying of the wetlands and the Japanese alder incursion, meaning that red-crowned cranes have been losing ideal habitat the whole time. Both populations are afforded advantages and disadvantages that could be said to cancel one another out. Yet the AWB whooping crane population today is much smaller than the red-crowned crane population—everyone is in agreement on that.

I told them that the best explanation for this variance is almost certainly the winter feeding campaign. Kawase and Matsumoto agreed. They're both familiar with the whooping crane's story but declined to speculate about why that recovery effort had hit so many snags as it had in the past. Still, they could say with some degree of certainty that the much faster recovery of the red-crowned crane is almost certainly attributable to winter feeding—by simply tossing corn at the red-crowned cranes for three months of the year, the Japanese conservation authorities achieved far faster population growth over the decades compared to the parallel whooping crane effort in America.

"However," Kawase retorted, "recently there's been talk that the population may actually be falling."

"Do you think this is because their feeding volumes are being reduced?" I asked her.

She nodded her head. "So is the experiment going well?" It was a rhetorical question. She had her answer. "It doesn't seem so."

The "experiment" Kawase alluded to is ongoing at this very moment, launched at the behest—no, the command—of Japan's Ministry of the Environment.

The federal government in Japan has long been concerned about the red-crowned crane's relatively sedentary ways. In short, they want their cranes to migrate as the ones in Siberia and North America do, or at least expand their range beyond Hokkaido's eastern fringes. Hokkaido's cranes do move about, probably more than most observers give them credit for. They can be found several miles to the east beyond the vicinity of Kushiro at the Kiritappu and Bekkanbeushi marshlands near the fishing town of Akkeshi and even further east in the region around Nemuro. Red-crowned cranes are also now a more or less permanent feature of portions of the Tokachi region near the city of Obihiro. A breeding pair has even been spotted near Tomakomai, just south of the prefectural capital of Sapporo. Most encouraging for conservationists, a lone migratory yet decidedly Japanese *tancho* has been seen by witnesses enjoying the vast open spaces of the Sarobetsu wetlands just south of Wakkanai, Japan's northernmost city.

But the Ministry of the Environment wants this species to spread its wings over a far wider swath of Japan. The fact that a breeding pair is known to reside in the area near Tomakomai suggests that this species will not stay confined to Hokkaido for long. Eventually, I believe, *tancho*'s numbers will rise to a point where this remarkable species will make a triumphant return to Japan's main island of Honshu. But a few years back, Japan's government thought that this wasn't happening quickly enough. The Ministry of the Environment hatched a plan to nudge *tancho* in that direction.

Beginning in late winter 2015, the ministry ordered the three feeding stations—Kawase's at Akan, the Itoh sanctuary, and Tsurumidai—to reduce the volume of feed (still mostly corn) served up to the red-crowned cranes by 10 percent. Subsequent reductions of 10 percent were to occur every subsequent feeding season—by 2016, feeding was to have been reduced by 20 percent, by 30 percent when winter feeding finished in 2017, and by 40 percent by the time I arrived in Hokkaido and began taking a closer look at the whooping crane's distant Asian cousin. When I met Kawase at her base in north Akan, the red-crowned cranes were being fed about half the feed in terms of weight compared to prior years. Back then, the ministry's stated plan was to eventually reduce feeding volumes to zero, suggesting that by 2025, *tancho* was supposed to be left completely to its own devices during the cold and snowy northern Japan winters.

The situation was getting tense back then, and I imagine it only intensified as the years rolled on. Farmers began complaining that the cranes were raiding their fields and even their cattle pens with greater frequency than in the past. A few cranes were accidentally killed by farm equipment. Matsumoto and Kawase both acknowledged that the seemingly rising number of anecdotes involving conflicts with humans and injured or killed cranes led them to believe that the ministry's strategy was backfiring badly.

But it's hard to know if they're right, especially because the folks responsible for population counts are rounding their population estimates by 50 birds. So, indeed, two back-to-back counts suggested that the population of red-crowned cranes residing in East Hokkaido had stalled at about 1,800 birds. The communities that had taken care of these birds for more than half a century began protesting, voicing loudly their concerns that doing what the Ministry of the Environment was telling them to do would not only harm the cranes, but it could eventually devastate the local economy, dependent as it is on red-crowned crane tourism, especially during the winter months. Villagers at a meeting in Tsurui told ministry officials point-blank that they planned to defy them completely, continuing with the feeding regimen regardless of official government orders.

Then came the pandemic.

COVID-19 seemed to put everything on hold in the region, including my research. Tourism certainly collapsed, leaving centers like Akan with far fewer resources to undertake its activities. The loud protests only grew louder. The Ministry of the Environment eventually relented.

In September 2020, at the height of the pandemic crisis, the government announced that it would ease its planned feeding reduction schedule given concerns over the impact to the red-crowned crane's population. Feeding volumes still would be reduced, the authorities insisted, but by 5 percent per season rather than the previous 10 percent annual drawdowns, extending the feeding regimen by at least five more years according to this updated plan.[1]

This move didn't satisfy anyone, however. So the ministry relented again, at least somewhat, as it continued to try to win the locals over to its way of thinking.

In late 2021, at a gathering at a community center in Tsurui (officials from Sapporo participated online), the Ministry of the Environment announced a new temporary adjustment to the feeding volume reduction plan. The volume of feed given to the cranes at the Akan center would continue to be cut, but the two feeding stations in Tsurui were given a reprieve—they'll be allowed to maintain feed volumes at 2020 levels for at least three more years.[2]

I have yet to ask Kawase-san what she makes of this new plan. I doubt she likes it. One argument she made was that it may not be practical for cranes to migrate to other parts of Japan anymore. Meaning, they may be stubbornly sticking to their current favorite stomping grounds for a reason. It's a jump for them to cross the sea from Kushiro to Iwate Prefecture on Honshu, let alone to find any suitable wetland habitat that may be there. Meanwhile, there's the rest of Hokkaido to figure out before the red-crowned cranes can discover more land across the Tsugaru Strait.

"In Hokkaido, it snows everywhere, and often they simply can't get food," she told me, meaning that the snows are too deep on much of the island for them to dig through to find any forage that might be had below. "They can't necessarily go anywhere else."

Cut off the winter feeding completely and then what? That's her main concern. "So where are they going to get food? Farms, ranch operations, taking food from livestock, and we're seeing losses in the population from accidents."

As Kawase told me this in a heated voice, Matsumoto nodded his head vigorously, like a teenage headbanger at a heavy metal concert. In his mind, it's simply too late for this particular endangered species management regime to turn back now. The red-crowned cranes had simply become too habituated to human aid, almost domesticated at this point, he argued. The land and habitat were limited, and the population was large. Letting some 2,000 cranes fend for themselves might work in a place like Texas, he said, but not Hokkaido, or anywhere else in Japan, for that matter. "If we had huge open spaces like in America, then it would be an easy thing to simply let them go off and live in those places," he told me. "In Japan, space is limited."

Here I was at the scene of one the world's most successful endangered species recovery stories, interviewing two top officials about their enormous success, and they were rejecting a central premise of my argument. Has East Hokkaido been more successful at crane species recovery than southeastern Texas? Absolutely. Has artificial feeding been key to this overwhelming success? Almost certainly, they believe; at least they can't think of any other plausible explanations—nor can I. Couldn't the powers that be in Texas learn a thing or two from Japan? Sure. But it's not that simple, they insisted.

Despite the struggles at Aransas, both Kawase and Matsumoto expressed to me their admiration for the AWB whooping crane recovery program. Envy, even.

Artificial feeding for red-crowned cranes was a grassroots, community-driven activity launched by Yamazaki in the Akan area in 1950 and later copied by Itoh on the edge of Tsurui (the two men almost certainly knew each other). Professional wildlife managers were powerless to stop them, or perhaps they were otherwise preoccupied with postwar recovery efforts. Rather, they later adopted and formalized the practice, instituting controls and statistical regularity to the systematic feeding. But Matsumoto and Kawase talked as if there's a general consensus today that this might have been a mistake, though they dare not say this too loudly or too publicly.

The official propaganda continues to celebrate figures like Itoh and Yamazaki as heroes. The artificial feeding is still considered to be at the root of the red-crowned crane's recovery and ongoing survival. Indeed, that species is now no longer considered technically "endangered" according to the Ministry of the Environment (MoE) and the International Union for Conservation of Nature. MoE may have anticipated this, leading it to order the feeding stations to start weaning the cranes off human support in the first place. But per Matsumoto and Kawase's argument, that should have occurred a long time ago if MoE was ever serious about seeing the red-crowned crane return to a truly wild, completely independent existence.

They both think the Aransas managers may have had it right all along.

Sure, it's taking a lot longer for the whooping crane to recover. That species is still endangered despite the Fish & Wildlife Service's recent attempts to move the goalposts closer than is likely advisable. But at least whoopers will remain wild and independent as their numbers grow, Matsumoto told me. As for *tancho*? It's probably too soon to say.

Matsumoto clarified for me that he appreciates what the ministry officials in Tokyo are trying to accomplish. But he was of the opinion that they should relent, especially considering his fears that the most ideal wetland habitat to be found in Kushiro Marsh was still being lost to the alder incursion. "Truthfully, I sincerely hope that the cranes widen their range as we lower the amount of food we give to them," he told me earnestly. "But in reality, so far they are keeping to this small area" of East Hokkaido.

Kawase Miyuki strikes me as a glass-half-full type of person, too. She took our summer of trespassing at the Akan International Crane Center in stride, kindly informing us of our inadvertent infraction and directing us back to the designated path. An overzealous park manager in the United States probably would have moved to prosecute us. Just as she gently nudged my wife and me back to the officially designated public areas of the park, Kawase seemed hopeful that she and her allies in this discussion, meaning the majority of wildlife practitioners living and working in East Hokkaido, eventually could nudge MoE leaders in Tokyo gently back to their way of thinking. As noted, I haven't had the opportunity to ask her yet, but I imagine the ministry's recent moves to ease restrictions on the feeding reduction program probably give her some hope even though Akan was left out of the most recently revised program.

Perhaps one day the red-crowned cranes of East Hokkaido will no longer require such active human intervention in their affairs, whether to maintain tourism or otherwise. But that day won't arrive anytime soon, Kawase firmly believes. "If humans are continuing to cause losses to the cranes' habitat," she argued, "then I think we probably shouldn't stop feeding them."

There's still time.

The government's original plan was to end feeding entirely by 2025 and to see how things go from there. But the conservation program pioneered by Yamazaki, Itoh, and others like them has won a stay of execution. Community members in Tsurui insist that they'll continue scattering corn to the cranes in some volume indefinitely, regardless of Tokyo's directives. Near as I can tell, folks at the Ministry of the Environment are pretending that they didn't hear this. For now, they're still in charge, but they've decided to slow things down, even pause matters in Tsurui, and then monitor the red-crowned crane's numbers for signs of either continued expansion or stagnation. But they'll mostly be looking for signs of migration. They very much want the rest of Japan to enjoy and take care of these remarkable, beautiful animals.

If the cranes are turning up in Tomakomai, then they almost certainly will end up finding their way to Aomori Prefecture due south. That's my bold prediction. And this event will make huge waves throughout Japan because, with the exception of zoos, *tancho* hasn't been seen outside the island of Hokkaido for more than a century. The Japanese might screw it up, though. According to one tale I heard, a few years back, a wayward *tancho* found its way to a winding river in a plain just beyond the outskirts of Sapporo, but an overzealous photographer apparently chased it into traffic, killing it. *Tancho*'s return (inevitable, I believe) to the northern tip of Honshu probably will be greeted with hundreds if not thousands of photographers and assorted tourists, but if the authorities play it right and these curious onlookers restrain themselves, then we might get lucky and MoE might get what it's been after since at least 2014.

In April 2022, I got the Red-Crowned Crane Conservancy's most recent population estimate. I wasn't very happy to see it. RCCC again revised its estimated number of Japanese cranes down to 1,800, from 1,900 at the close of the prior survey year. But I can't completely trust this figure, given its penchant for guesstimation and rounding by units of 50, and neither should you. If they're right, then the red-crowned crane population has stopped expanding in Hokkaido, conspicuously when MoE began manipulating winter feeding volumes, lending further credence to my suspicion that the winter feeding has been the main driving force behind this previously endangered species' recovery. I suspect the census count is wrong, but I can't make that call definitively. I say this only because Hokkaido is vast—it's Japan's second largest island—yet relatively sparsely populated. About 5.2 million people live there, with maybe half or fewer of all Hokkaido residents concentrated in or near Sapporo. Hikers still occasionally and tragically get lost in Hokkaido's vast expanses of wilderness. There are plenty of places for clever cranes to hide, especially in the northern stretches of the island.

I trust Kawase Miyuki's estimate—there are probably about 1,900 to 2,000 red-crowned cranes in Hokkaido today, and likely more by the time this book is published. At least, that's my hope. The AWB whooping crane population probably still will be fewer than 600 cranes by then, though I would be very pleased if the folks at Aransas announce that whoopers at the wildlife refuge crossed the 600 milestone for the first time. I wish both programs success, though up until now, the effort in Japan has clearly enjoyed more of it.

I expect further changes to Japan's Ministry of the Environment's plans depending on the crane's population trajectory and how loudly the citizens of Akan and Tsururi complain. But barring any major changes to trends that have been underway for decades now, by 2035 the AWB whooper population should be at or more than 1,000 cranes, meeting the goal of Canadian conservation authorities. But the red-crowned cranes will be at least double or, more likely, closer to triple that figure, especially if more cranes find the region of Tomakomai and Sarobetsu as hospitable as the early pioneers to these frontiers do. This despite the fact that the red-crowned crane has never enjoyed the protection of the Endangered Species Act (not directly, anyway). This despite the fact that red-crowned crane recovery was not launched or originally led by professional wildlife experts with doctoral degrees operating in a top-down manner, but rather by unsophisticated yet concerned individuals and laypeople willing to experiment with unproven techniques in endangered species conservation and population recovery.

Still, we need to be careful here. We should expect the unexpected. We should avoid jumping to conclusions.

There are many lessons to take away from 50 years of the Endangered Species Act's existence. Many of those lessons can be derived from well outside the jurisdiction of the ESA itself, including on the other side of the Pacific Ocean. But past isn't always precedent, and circumstances can and will change. As best as I can tell from this tale of two cranes, feeding AWB whooping cranes during periods of acute food scarcity—for example, during nasty droughts—could result in more whooping cranes surviving and thriving, especially fledglings, leading to more reproduction in the population and thus faster population growth.

After years of investigating this tale of two cranes, the conclusion I keep running into is that the winter feeding program is the force behind the variance in these similar endangered species' recovery rates. In short, the population of red-crowned cranes rebounded much faster than the whooping cranes because the Japanese intervened to ensure the cranes had enough food during periods of low forage availability, while the authorities at Aransas National Wildlife Refuge did not.

The AWB whooping crane endures a very long and taxing migration to and from Canada and Texas twice a year. Yet there is little indication that the AWB whooping crane population experiences relatively higher rates of mortality during this migration, and I've yet to hear that the crane's numbers are being limited by higher rates of predation during their time in Wood Buffalo National Park.

We certainly can rule out habitat availability as a factor in the cranes' different rates of reproduction success. AWB whoopers have far more habitat available to them, both for breeding and winter feeding. This is especially true when you factor in the agricultural land adjacent to the Aransas refuge, where whooping cranes are known to forage (as I've witnessed myself). Red-crowned cranes forage in the farmlands next to Kushiro Shitsugen, just as the whooping cranes feed at the farms surrounding the Aransas refuge. But from my very basic investigations using satellite imagery and GIS software, I can say that the AWB whooping cranes have at least twice the habitat available to them near Aransas as the red-crowned cranes within the same radius. Not to mention that as the red-crowned crane's numbers have risen, the actual area of ideal habitat available to them within Kushiro Shitsugen has been shrinking due to the siltation and Japanese alder incursion problems I described earlier.

Whooper has a tough commute back and forth from Canada to Texas. *Tancho*, meanwhile, has to contend with very mean winters. Whooper should be feeling far less stress from human presence, as well. Aransas is surrounded mainly by small towns. The nearest urban core near Aransas is Corpus Christi, with a metro population of about 400,000, which is about a 90-minute drive away. You can literally walk to Kushiro Shitsugen from Kushiro proper, with the city and town combining to comprise a total population of about 200,000 residents. And based on my experiences, I think it's safe to say that the region of East Hokkaido where the red-crowned cranes are found draws far more tourists annually than Aransas, at least before the coronavirus pandemic.

Both species are afforded advantages and disadvantages that more or less cancel each other out. All that's left is the one advantage *tancho* enjoys that whooper does not: food made available artificially during times of natural food shortages.

This gets to the heart of my thesis. You recall how I explained that the red-crowned crane and whooping crane display similar or even near-identical reproductive cycles. They both have similar lifespans and boast basically the same biotic potential. Females of both species usually lay two eggs at a time, and both eggs will hatch if spared from predation, though only one adolescent usually survives and makes it to adulthood. In Aransas, they tried to tip the scales in favor of survival by taking one egg from nests and rearing those chicks in captivity. But they of course can't do this for every nest. In Hokkaido, however, I think they sort of did.

I suspect that the winter feeding program increased the odds that the red-crowned crane's second fledglings survived to maturity, and throughout the entire population, not just for a small number of chicks hatched in captivity. Because of the secure food supply (even if the feeding stations are not the cranes' only source of winter food, which they aren't) both parents remained well fed and healthy enough to give birth to stronger and healthier offspring. These younger cranes, in turn, were able to survive the winter with better odds thanks to the artificial feeding campaign. Thus, a relatively higher proportion of second eggs in the red-crowned crane population survived compared to the AWB whooping crane population. These survivors gave birth to their chicks, and on and on, resulting in a faster overall rate of population growth. Or at least, that's how I suspect things have been playing out since at least 1950.

Meanwhile, reproduction for AWB whooping cranes stayed at its relatively lower natural rates, despite some success with the Fish & Wildlife Service's captive breeding program. Then droughts entered the picture, tossing a curveball at the entire operation. More young cranes starved to death, leaving fewer to give rise to the next generation. Thus, while whooping crane numbers have steadily increased, red-crowned crane numbers have risen faster. Whether they're still rising is an open question, but the past data is clear.

So, yes, I think introducing artificial feeding to AWB whooping cranes during periods of acute food shortages could facilitate faster population growth, at least for the AWB whooping cranes, the focus of this book. However, "could" and "should" are two very different things. Just because some activity could net positive results, that doesn't necessarily mean that's the precise activity you should pursue. As Kawase and Matsumoto taught me in Akan, the complete picture is never quite that simple. Something works until it stops working. That same something could solve one problem today, only to create other problems down the road.

As Ingeman and colleagues argue, the global extinction crisis is worsening to the point where conservationists need to seriously consider what works and what doesn't, to zero in on case studies of successful efforts of saving species from the brink of extinction to see what clues can be gleaned from these most successful endangered species recovery efforts. We should figure out what saved the peregrine falcon, and why it worked.[3] The roseate spoonbill was nearly annihilated from North America entirely, but today it's a common sight along the Gulf of Mexico coastline. We might benefit from fully understanding how. The Japanese serow—the cross between a deer and a goat that I alluded to earlier—was nearly wiped out, but authorities stepped in to save it, and today this strange creature is listed by the IUCN as a species of "least concern," likely abundant enough to sustain hunting pressure should the need or desire arise. How? Why? We should find out. The same goes for the red-crowned crane. I think I already know the answer, but other researchers far more skilled than I might want to take a stab at this question anyway.

Lessons from the past absolutely should be considered and taken seriously. But let's not get carried away. Caution is in order, especially if we're proposing to simply take these past lessons and apply them to the present, expecting similar or near-identical outcomes by extrapolating into the future.

Sometimes it appears, at least from my estimation, that artificial feeding might very well be the "right call" when faced with a struggling species on the verge of extinction. It ultimately may have saved Florida's manatee, if not the South China Sea's dugong. But for other endangered species, when is it the right time to make that "right call"? Is there ever a right time to make such a call? The professionals in Japan didn't make that call for *tancho*—two dairy farmers did. Would the pros have acted differently if given the chance?

CHAPTER NINE

~

Leopold Revisited, Again

So, fine, just because artificial feeding *could* lead to much faster endangered species population recovery doesn't necessarily mean one *should* employ this specific recovery strategy.

But what if you did so anyway?

The academic literature on artificial feeding for species recovery is rather mixed—some studies show that this strategy could end up harming a species' population in certain circumstances—but there is compelling evidence emerging from the research community that intervention feeding is likely a powerful tool for recovering endangered avian species. In other words, artificial feeding could be for the birds. The Japanese may have been the first to discover this, but researchers across the world now are reaching the same basic conclusion to varying degrees.

Of course, the jury is still out, but let's examine a small sampling of the evidence in support of supplemental feeding.

First, it's worthwhile to clarify that intervention feeding in the wild isn't all that rare. As I mentioned earlier, the U.S. Fish & Wildlife Service and Florida's state conservation officials joined forces to save manatees from starvation during the 2021–2022 winter by dumping large volumes of lettuce in the manatees' home waters. Authorities occasionally drop bales of hay in Yellowstone National Park to help out bison and other herbivores struggling to find forage after particularly heavy snowfalls. FWS even tried feeding whooping cranes briefly in the 1960s. And in fact, millions of Americans artificially feed birds with back yard feeders. I'm sure there are numerous other examples that American conservation authorities and experts could highlight for me. The main difference here is most or nearly all of these interventions were short-term affairs. They were meant to be temporary measures taken only in times of potential emergency (the 1960s whooping crane feeding seems to have been mainly a quickly abandoned experiment). There may be a case out there somewhere, but I have yet to identify a sustained effort in which FWS or other U.S. authorities leave food for endangered animals or animals facing extinction in the wild on regular intervals spanning several years, let alone decades. Leopold's philosophy is still paramount—do not feed the wild animals (unless absolutely necessary and only as directed by designated authorities, of course). In other parts of the world, people seem more willing to experiment with regular and sustained long-term artificial feeding.

To avoid uncomfortable conversations about sex, Westerners used to have a novel and peculiar explanation for children who ask where babies came from. I'm sure most who are reading these words know the traditional, conservative, and fantastical answer/lie given to inquisitive children in these circumstances: storks deliver babies to couples.

The legend of white storks delivering babies likely originated in Europe during the Middle Ages. They were a common sight back then. As I've read, white storks were fond of building their nests on people's rooftops. This probably made the stork explanation somewhat believable to medieval children (though I like to imagine that most weren't that credulous, but were humoring their parents). Less legendary is the tale of how the white stork was nearly driven to extinction in Europe, especially after World War II. Conservationists and communities in Europe struggled mightily to bring the white stork back from the brink. Captive breeding and reintroduction programs started in the 1950s in some countries. Strong laws were enacted to protect the species. Despite all this, for decades, the stork's recovery resembled in many ways the fits-and-starts, forward-and-backward progression of the Attwater's prairie chicken restoration initiative. As Hilgartner and colleagues recount, only 11 breeding pairs could be found in all of France in 1974, 5 breeding pairs in the entirety of the Netherlands in 1984, and 6 breeding pairs in Denmark in 1996. Since the 1950s, conservationists have thrown everything and the kitchen sink at saving this species, including "continuous releases of white storks reared in captivity, installation of nest sites (poles), and supplementary feeding of free flying individuals."[1] Their efforts seem to be finally paying off; the most recent white stork population censuses taken in Europe show an upward trajectory.

In their 2014 study, Hilgartner and colleagues report multiple reasons for the white stork's recovery. This species is migratory, with most storks preferring to spend winters in West Africa rather than Europe, so fewer and milder winter droughts are seen as one possible explanation for the birds' recovery. Scientists also think the storks have learned to thrive on the presence of a certain invasive species—in this instance, the Louisiana crawfish now abundant in many of Europe's wetlands (another case in which intervention to restore a habitat to its native state actually could end up harming an endangered species). But researchers behind this study sensed something missing from the academic literature on the white stork. "To our knowledge, the effect of additional feeding on reproductive success of white storks has never been investigated in detail," they wrote.[2] Or at least until they decided to do just that.

As with the case of the red-crowned crane, fans of the white stork in Europe established feeding stations for the birds in various locations. These feeding stations are probably found scattered all over Europe, but the authors of this study focused on a set of feeding sites established in southwestern Germany near the border with Switzerland to determine if there was any correlation between breeding pair reproductive success and proximity to feeding stations. The data they collected spanned 22 years, from 1990 to 2012, including 569 breeding events at 80 separate locations. Here's another example of a sustained artificial feeding campaign focused on one endangered avian species lasting decades, though this study used only two decades' worth of data. Lo and behold, this fairy tale ended with a twist—through artificial feeding, the people delivered more babies to the stork, in a roundabout sort of way.

"Reproductive success was significantly higher in pairs breeding in close distance to the feeding site," the authors concluded. The correlation is strong enough to tell the researchers that there's definitely some effect. The investigators found that the number of fledglings per nest declined by 8 percent for every kilometer away from a feeding station. In short, white stork couples living in closer proximity to feeding sites had overall better reproductive success than those breeding pairs nesting further away. This research team seems convinced—they argue that artificial feeding seems to be not only a good way to boost reproductive health in an endangered species (or at least for endangered storks), but this intervention method probably also compensates for a relative lack of ideal foraging habitat available in the vicinity. "Supplemental feeding of free flying white storks as a conservation measure should therefore compensate for low habitat quality and increase reproductive success of pairs," they wrote.[3] In other words, in the absence of sufficient habitat area or quality, toss the birds some food if you want them to survive, thrive, and multiply.

These findings offer additional support for my hypothesis, at least on the surface. In the wild, animals have two goals: survival and reproduction. All things being equal, survival takes precedence; reproduction is secondary. If an animal's basic survival needs are satisfied, then more energy and effort are spent on reproduction. If not, the animal, under elevated stress, focuses more energy and effort on its own survival. Through artificial feeding, humans take over and effectively relieve the stress. This source of anxiety is more or less eliminated, freeing the animals to focus more effort and energy on reproduction and the successful rearing of their young. Thus, in endangered species rehabilitation initiatives, conservation authorities may achieve faster population growth in a target species through concerted, long-term artificial feeding, especially at times when natural forage conditions are less than ideal.

For a species on the verge of extinction, options to recover the species include population management, habitat management, or a combination of both. Population management methods include activities such as captive breeding and artificial feeding during periods of low food availability (going by my definition here). Habitat management means legal restrictions on human activity in the species' habitat and then maintaining that habitat. Japan has done both since the 1950s vis-à-vis the red-crowned crane. The Europeans, it appears, have been doing the same thing with respect to the white stork. For the AWB whooping crane, the United States largely ended population management in the 1960s, with the exception of some captive breeding activities and health interventions, and shifted overwhelmingly to habitat management, a stance that has dominated whooping crane management for several decades now. This difference probably best explains why Japan's program has shown more success relative to the U.S. AWB whooping crane effort. This fascinating study of white stork fertility success in proximity to artificial feeding stations further supports this view.

There's other evidence, some of it indirect.

The island of Mauritius is famous as the former home of the dodo, that lovable, goofy-looking flightless bird that was wiped off the face of Earth by careless European visitors during the course of only a few decades. That tropical paradise is home to other struggling species, including the Mauritius parakeet, today a critically endangered species. The species survives thanks to captive breeding programs, reintroduction to the wild, habitat conservation, and, it seems, supplemental artificial feeding in the wild. Managers at Aransas once told me that they opted against artificial feeding for the whooping cranes because they're concerned that concentrating cranes into feeding sites could encourage diseases to spread through the population. One could point to the red-crowned crane in Japan as evidence that such worries are likely overblown. It now appears the Mauritius parakeet provides another case study in support of artificial feeding for endangered species recovery and survival.

University of Kent researchers shared the same concerns held by Aransas managers—they feared that artificial feeding facilitated disease transmission in some populations of Mauritius parakeets. So they gathered about 20 years' worth of data on the population of a particular group of parakeets to see how well the birds were recovering. They were looking for signs that the feeding stations were detrimental to the parakeets' recovery, conversely causing higher incidence of mortality by making it easier for diseases to spread in a small population of reintroduced birds. Their fears were well-founded. "Infectious diseases are widely recognized to have a substantial impact on wildlife populations," as Tollington's research team explained in their study, published in 2015 in the *Journal of Animal Ecology*. "These impacts are sometimes exacerbated in small endangered populations, and therefore, the success of conservation reintroductions to aid the recovery of such species can be seriously threatened by outbreaks of infectious disease."[4] Seems logical enough. However, what they actually discovered surprised them, the authors wrote.

For sure, they found some association between supplemental feeding and disease spread, in this case, beak and feather disease virus (BFDV)—birds that gathered in groups and relied more on feeding stations tended to show higher incidence of the disease. But the difference was insignificant. Other populations of parakeets on the island were far less affected by these disease outbreaks, they found, reporting that the outbreaks tended to stay concentrated among the birds heavily using supplemental feeding stations. But these outbreaks didn't last long. And whatever negative impact on the parakeet population that the diseases may have had was entirely overwhelmed by the positive impact of the artificial feeding regimen.

They bury the lede in their study somewhat, but the researchers discovered the same general effect that has been discovered with the white storks in Germany. As they clarify later in their study, "Parakeets which took supplemental food generally fledged a higher proportion of chicks than pairs which did not use this resource." Mauritius parakeets taking advantage of feeding stations gave birth to and raised more chicks relative to members of the same species on the same island that were not relying on supplemental feed. Why? Likely, the researchers found, because occasionally these birds encounter periods of relative resource scarcity in their environment, making survival more difficult. The exception is the birds that had this source of anxiety largely eliminated by the human-provided supplemental food. BFDV is dangerous, especially for younger birds, but the artificial feeding bolstered the conservationists' reintroduction efforts, resulting in a remarkably resilient endangered species population recovery. Their conclusion: "Supplemental feeding, therefore, appears to be beneficial by (i) enabling breeding pairs to fledge more offspring and (ii) by mitigating the negative effects of fluctuating natural food resources."[5] They were expecting to find the greater disease communicability to be a significant problem in the Mauritius parakeet's recovery, but instead found it to be a non-issue.

Other scientists have found that supplemental feeding seems to facilitate greater survival and reproduction in species of avian scavengers—namely, vultures and condors.

This practice poses risks, for sure. I've read of cases in which wildlife managers found evidence that feeding stations established for particular prey species tend to benefit their predators more—the predators learn to lie in wait for animals visiting feeding stations, making their hunting work easier. It seems reasonable to expect that concentrating animals at feeding stations makes it easier for diseases to spread. Parasites, too, for that matter. Researchers have long recognized the negative impact on population growth trends caused by higher animal population densities. In short, animal populations are more exposed to limiting factors like predation, diseases, food shortages, and so on at higher population densities. But there are clear benefits, as well. The question is, do the benefits outweigh the risks in other cases, as appears to be the case for the Mauritius parakeet?

Vultures and condors have been hit hard by ranching pushing out native grazing species. In the wild, buffalo, deer, antelopes, and so on die, and their carcasses stay right where that animal dropped. This is the carrion that avian scavengers rely on for survival, in turn playing an important role in the ecology by breaking these carcasses down and returning their energy and resources to nature. Cows, by contrast, mainly die in slaughterhouses, unless predators take them in the wild (and ranchers have largely eliminated these predators). Thus, fewer ungulates means less available carrion, decimating populations of flying scavengers in particular, as this is pretty much their only source of food.

Globally, people decided to save vultures and condors from extinction by leaving food for them, without fully understanding the implications of doing so, much like the case of the folks who first began to feed the red-crowned crane. Folks in Akan and Tsurui simply wanted to save the birds from oblivion; they didn't know what other effects might arise. In a 2016 edition of *Frontiers in Ecology and the Environment*, a team of researchers cover both the pros and cons of artificially feeding avian scavengers in detail. Studies on species of vultures in Egypt and southern France linked artificial feeding to endangered species population recoveries. Some even speculate that artificial feed can save vultures by keeping them away from the carcasses of animals that have been poisoned. And there is additional, if controversial, evidence that populations of vultures and condors enjoying artificial feeding interventions enjoy greater reproductive success in turn.[6] The picture is mixed, for sure, and much more research is necessary, but this generally mirrors findings from the stork and parakeet studies.

Some have even modeled the likely effect of artificial feeding on species of vultures, predicting mainly positive outcomes, though not completely discounting the risks.

In 2014, Ferrer and his research team set out to determine how a concentrated artificial feeding program might benefit or harm populations of endangered bearded vultures in Spain. They proposed using supplementary feeding to boost populations of vultures living in suboptimal locations and then introducing these extra individual vultures to other regions where the ecosystem was more amenable to survival without human interventions. It's a clever idea, sort of like captive breeding but in the wild. Their modeling shows that it should work. "Using food supplementation in target territories, the expected production of extra young allowed their removal without any effect on the donor population, in either the short or long term," the study concludes. The researchers figure the best results will come from a sustained effort lasting more than a decade, or "at least 13 years," they wrote. Combining artificial feeding and translocations is also substantially cheaper than captive breeding programs, the research team points out. "Analyses showed that the annual budget of a captive breeding program for this species could be seven times more expensive than the translocation of extra young produced by food supplementation."[7] This is worth repeating because I want you to remember this for later: according to these researchers, the cost of breeding endangered species in captivity and then releasing them to the wild is up to seven times higher than the cost of simply feeding those same animals in the wild and allowing their numbers to increase that way.

So far, I've deliberately kept this discussion focused on bird species.

Through my research, I've become convinced that sustained artificial feeding during periods of scarce forage availability almost certainly resulted in higher rates of reproduction for red-crowned cranes. The fact that the red-crowned cranes' population recovery seems to have hit a snag now that officials are trying to wind down the feeding campaign suggests that I'm likely correct. I'm a bit more optimistic than Kawase and her colleagues. I think the red-crowned crane population will prove resilient and spread its range across a wider swath of Hokkaido and eventually to Honshu, as recent trends suggest. Hilgartner and colleagues find convincing evidence that white storks using supplemental feeding resources reproduce at faster rates than white storks that do not. Tollington and others express their surprise to find that not only do artificially fed Mauritius parakeet populations reproduce at faster rates than parakeets relying entirely on wild forage, but that the availability of supplemental feed equates to a high degree of endangered species population resilience during the recovery period. Vultures and condors aided with supplemental food not only seem to enjoy better chances of survival, but they may also give birth to more fledglings than might otherwise be possible, and this effect could probably be leveraged to boost populations in other regions. It's not definitive, for sure, but the available evidence in support of population management through sustained, long-term artificial feeding is rather compelling.

This method of endangered species management probably works quite well for some species that don't fly. For example, a study published in 2013 by the Czech University of Life Sciences looked at the impact of artificial feeding on the western giant eland of West Africa. You recall that I argued that artificial feeding likely alleviates animals of the stress and anxiety associated with the struggle to survive during lean food periods, enabling them to focus more energy and effort on reproduction and rearing their young. These scientists see the same forces at work for the eland, a beautiful animal that looks something like a cross between a zebra and an antelope. There certainly are risks, like encouraging less healthy eating habits in the population, but these appear to be outweighed by the benefits. "Supplemental food did not induce changes in browsing pattern at the plant species level, probably due to a small individual effect on total nutrient energy intake," Hejcmanová and her coauthors wrote. "Food supplementation, however, facilitates the animals overcoming unfavorable conditions or alleviates stress with additional rest, and could therefore assist as a conservation intervention to enhance fitness."[8] I bet this method also helps these less-stressed elands to give birth to and raise more baby elands in the future.

There are cons, for sure, don't get me wrong.

One of my former graduate advisers argued that it's possible for a sustained supplemental feeding effort, even a program conducted for just part of the year, to get carried away. A species' population could expand beyond the natural carrying capacity of that habitat, she explained. That could leave wildlife managers in a rather sticky conundrum. She makes a good point. Remove the food, and you may see a sudden and dramatic population collapse, one sure to be noticed by the broader public and media. Maintain the feeding, and the animal population may continue to expand beyond a level that's considered naturally sustainable. And we may not know exactly where that threshold lies. Ecosystem carrying capacity is a sound concept, but it's often hard to clearly define. We may not know exactly what a region's carrying capacity is, even if we do know the relative availability of resources there capable of sustaining a population. That's because many animals, like humans, change their survival strategies and shift to alternative resources when their preferred natural sustenance becomes scarce. But if conservationists decide to lead the recovery of an endangered species primarily through an artificial feeding program, they could find themselves somewhat trapped, unable to reduce or even eventually end a feeding program as easily as they may want to. Careful preplanning is probably in order.

Supplemental feeding may result in accidentally rendering a wild species quasi-domesticated. This is Kawase's fear—that the red-crowned cranes have become too dependent and expectant of the seasonal winter feeding from humans. They're certainly more acclimated to the presence of humans than the shy and reticent whooping crane. Personally, I suspect these fears are overblown. *Tancho* seems to have retained a healthy skepticism of humans, near as I can tell, careful to not get too close to the throngs of photographers mobbing the feeding stations north of Kushiro every winter. And the feeding stations are open only three months of the year. But red-crowned cranes may have come to strongly associate humans with food, especially in winter. That can't be a good thing, even in a post-Leopold conservation world. And, as mentioned earlier, I've read about cases in which the establishment of feeding stations to help some species is suspected of inadvertently helping those species' predators more. The predator animals get clever and learn to wait for their prey to check in at a feeding station, almost like we humans are baiting the animals for the carnivores.

Now, let's consider the pros.

Suppose feeding is limited only to times when natural foods are in low supply or not available entirely, like in a drought or during the winter. Supplemental feed clearly helps endangered species to survive these challenging conditions. Oftentimes, artificial feeding is continued, at least temporarily, even at the close of a round of captive breeding. The California condor was once rendered extinct in the wild because conservation authorities took the few surviving individuals into captivity to protect them and for captive breeding purposes. After the condors were rereleased into the wild, the authorities continued distributing food to the condors for about two or three years to ensure the birds' survival.

Now, there's solid evidence in the academic literature that artificial feeding not only facilitates individual and group animal survival, but also probably helps increase a population's rate of reproduction. Less energy exerted on individual survival means more energy left for other natural imperatives. What are the two principal things that drive life on Earth? Survival and, usually, reproduction (at least to the point when an animal population becomes too crowded). Healthy, well-fed adults give birth to healthy offspring. Greater food availability ensures that the offspring survive and thrive, growing big and strong until they create offspring of their own.

It's also a hell of a lot cheaper than captive breeding, which explains why the communities in East Hokkaido have been able to sustain their campaign continuously and uninterrupted since 1950.

By some accounts, it can cost more than $100,000 to hatch a whooper egg, raise a single chick in captivity, and then release it to its natural wild habitat. You have to pay experts to do this sort of work, and then pay for the facilities to incubate and house the chicks and adolescent cranes. Recent anecdotal evidence suggests that a captive breeding program is even more vulnerable as a sustainable policy tool than feeding stations. Whatever their motivations, folks at Japan's Ministry of the Environment have never said that they need to wind down artificial feeding because it is prohibitively expensive. It isn't necessarily cheap, but the feed and funds are easy to acquire (much of it can be donated). By contrast, a whooping crane captive breeding program easily costs millions of dollars annually. Someone in the Trump administration eventually caught wind of this, and a federally funded effort was shut down in 2017—those whooping cranes ended up scattered to zoos throughout North America. During times of huge government deficits and ever tighter budgets, it gets rather difficult to convince average folks and lawmakers that spending $100,000 per year on a single whooping crane—rather than some worthy kid's college tuition—makes sense. By contrast, I don't imagine it's all that difficult to find funds to cover just three months' worth of feed for hundreds or even thousands of individual birds or to convince the public to pay for such an effort; $100,000 buys a lot of corn in Japan. Trust me.

In late 2016, 18 conservationists came together to pen an article for the Ecological Society of America for the society's *Issues in Ecology* series of reports. Their 28-page overview is titled "Species Recovery in the United States: Increasing the Effectiveness of the Endangered Species Act." This is by no means the only critical take on the 1973 ESA, but it is among the most recent and better independent inspections of America's landmark wildlife and wild plant preservation law that I've come across. It came out shortly after I started my graduate studies in ecosystem science and management at Texas A&M University and my investigation into this tale of two cranes. This Ecological Society report partly inspired me to write this book.

I encourage you all to read it, as it covers more explicit policy territory than I have and with far fewer pages. But it doesn't cover everything.

The report begins with an overview of the history of the Endangered Species Act, an important reminder that the ESA does not mark the first time that lawmakers drafted special protections for wild plants and animals (and it won't be the last). The authors trace the ESA's roots to the 1900 Lacey Act, which preceded the 1918 Migratory Bird Treaty Act and the 1929 Migratory Bird Conservation Act. The federal government was granted the money and authority to acquire land as protected habitat thanks to the 1965 Land and Water Conservation Fund Act, reflecting conservationists' early recognition that preserving habitat would be just as critical to species' survival as controlling hunting and trade. The Endangered Species Preservation Act followed in 1966. Further tweaks and tightening of rules eventually led to the 1973 Endangered Species Act, which the authors see as "arguably the strongest piece of environmental legislation in history."[9] The ESA is also among the most internationally influential environmental laws, as I demonstrated earlier.

This particular report points out several things that are wrong with or lacking in how the ESA is implemented, though not necessarily with how the law was drafted. After all, implementation is where the rubber hits the road. As is the case today, in 2016 the ESA was credited with saving countless creatures and plants from extinction, but there were (and are) still far too many ESA-listed species experiencing only modest or struggling recoveries. As Evans and his coauthors lamented, "recovery for most species officially protected by the ESA—i.e., listed species—has been harder to achieve than initially envisioned." Reading these words again, I'm reminded of the Attwater's prairie chicken.

We can put other issues identified in their 2016 overview in the present and not the past tense. Funding for ESA protections can be inconsistent, and the funds that are received are unequally or inefficiently distributed. Something like 5 percent of listed species receive 80 percent of conservation and recovery money. Most funds are spent on fish, a strong indication that some clever congressmen found ways to use the ESA to subsidize recreational or commercial angling in their districts. Many species are threatened or endangered but are not so listed—Evans and colleagues put that number at probably ten times the actually number of listed species. And the threat posed by climate change hasn't been fully factored in yet. Their report was published not long after the Paris climate accord was adopted, so naturally they gave a nod to the ever-worsening global warming problem. A huge problem, they point out, is that a host of ESA-listed species experience such fitful and unencouraging recovery trajectories that they'll probably "require ongoing management for the foreseeable future," with the whooping crane perhaps a rare exception here, should the Canadian goal of 1,000 AWB whoopers be reached.[10]

Evans and his coauthors outline six "strategies" for enhancing America's endangered species management regime. They argue in favor of setting up a system for systematically determining funding priorities. More collaboration among federal agencies, nongovernmental organizations, and landowners is in order, they say. They also call for better species monitoring and "adaptive management" (that flexibility that I earlier called for), developing better and measurable endangered species recovery criteria, "climate smart" strategies, and more efficient "ecosystem-based approaches" to endangered species recovery. They also admit that there's a need for different attitudes in endangered species management regimes. Pretty and iconic species often are prioritized over less visually appealing species that actually may be in greater peril. For example, the bald eagle has seen a 25-fold recovery in breeding pair numbers since the 1960s. During the same period, estimates suggest that the Houston toad's population has declined by 75 percent or more. And although there are many iconic and rightly celebrated ESA success stories, they're too few and far between. At the time that Evans and colleagues wrote their overview, they estimated that less than 2 percent of all endangered species ever listed under the ESA recovered in population size to the point at which they could be delisted. But again, the law is only 50 years old, so more patience is perhaps required here. As I showed you earlier, there's good research evidence that an endangered species is more likely to grow in numbers and tend toward recovery the longer it's listed. Evans and colleagues concur here.

They note the most common endangered species management approaches and say that these can be improved with better science and a firmer understanding than what was available when the Endangered Species Act initially passed. The most common practices in the United States include habitat management and conservation, efforts to control invasive species, minimizing human impacts like pollution or poaching, and encouraging greater gene flow and genetic variety in an endangered species population through translocation or captive breeding and wild release.

They promote other relatively newer ideas. A major cause of the extinction crisis is that humans have broken up the land for their own uses, leaving animals and plants to survive on disjointed, disconnected islands of habitat and resource patches. We know that biodiversity is generally lower on actual islands—the kind in the ocean—and species endemic to islands are usually more vulnerable to extinction. Well, we've created this island situation on the continents. We need interconnections through either wildlife corridors or even vegetated bridges to help animals overcome obstacles like freeways and railroads. Better habitat connectivity helps to alleviate the disastrous situation our species has caused the others on this planet. Evans and his coauthors promote all this, as well as "assisted colonization"—helping species colonize places beyond their natural ranges where climate change may result in an environment becoming more suitable to an endangered species. In other words, perhaps in certain situations climate change can be used to conservation's advantage. It's an intriguing idea.

They also endorse a "coarse filter" approach to endangered species management, which is a fancy way of saying the Fish & Wildlife Service should lump several threatened and endangered species existing in the same region or range and develop a conservation strategy for saving them all as a group, not just on a species-by-species basis. FWS is already trying this in Hawaii, they note. "On the island of Kauai in Hawaii, FWS has grouped 48 listed species based on common factors such as shared threats and types of habitat, focusing restoration broadly on habitats and ecosystems rather than solely on individual species in their last few known locations."[11] Sounds like a more cost-efficient approach to me.

There's a lot of good stuff in this report. You should read it. But do you know what they left out entirely? Supplemental feeding for endangered species sustained over long periods, or at least artificial intervention feeding in times of natural forage scarcity. Those ideas don't appear anywhere in their discussion.

Perhaps this is an inadvertent oversight. Perhaps it reflects the traditional Leopold-era thinking in American conservation circles. But new thinking might be in order here, new flexibility in what we consider to be acceptable and unacceptable norms in wildlife management and conservation. Maybe we need a greater willingness to experiment, especially a willingness to learn from the abundant lessons provided for us by conservation stories outside the United States and North America.

Like those found in Mauritius.

Or in southwestern Germany.

Spain, even.

Or in East Hokkaido, Japan.

After all, as Ingeman and colleagues urged in their 2022 paper, "in the face of an accelerating extinction crisis, scientists must draw insights from successful conservation interventions to uncover promising strategies for reversing broader declines."[12] There are plenty of endangered species recovery success stories outside the United States. If only we'd pay more attention to them and, most critically, have the courage to learn from and emulate their examples.

CHAPTER TEN

Life Finds a Way?

How did life originate on Earth in the first place? Where did all this quickly vanishing biodiversity come from?

Honestly, no one knows for sure. There are several fascinating theories. One holds that life likely originated at the bottom of the ocean in hydrothermal vents. At those points found along deep ocean ridges where continental plates meet, extremely hot, pressurized, chemical- and nutrient-rich waters spew out of the ocean floor, creating chimneys. Unique ecosystems have evolved at these isolated points, the only spots on Earth where the food chain doesn't rely on photosynthesis but rather a process called chemosynthesis. Through photosynthesis, the root energy source underpinning life is sunlight; plants have evolved to utilize sunlight and water to make their own food, creating a food source for the rest of us—something eats the plant, then something eats the something that eats the plant, then something eats the something that eats the something that eats the plant, and on and on.

With chemosynthesis, bacteria have evolved to utilize the energy from hydrogen sulfide. Other critters come in to eat the bacteria. Then other organisms move in to eat the critters that eat the bacteria, and on and on. It's way more complicated than this, for sure, but that's the gist of it. To me, that chemosynthesis even exists is the surest sign yet that our universe absolutely teems with life. Even for planets that don't find themselves in the so-called "Goldilocks" zone—the right distance from the right kind of star—so long as those planets have water and enough internal pressure to generate sufficient heat, then life almost certainly will emerge at some point. Chemosynthesis could even be underway now somewhere else in our solar system.

Did life on Earth first emerge in the deep oceans with chemosynthesis, then later evolve to take advantage of the energy from the sun? Maybe. For now, the question stands: where does Earth's biodiversity come from? Here's one explanation, starting at the very beginning: the Big Bang.

What follows is a quick rundown of how it all unfolded as recounted by astrophysicist Neil deGrasse Tyson and his colleague Donald Goldsmith in the introductory pages of their popular book *Origins: Fourteen Billion Years of Cosmic Evolution.*[1] But I urge you to look up the details for yourselves from authorities on this topic, folks with the proper credentials and academic training that enable them to teach this subject. I don't necessarily qualify, which is why I lean so heavily on Goldsmith and Tyson, though they may not like the way I recount their fascinating and vivid explanation of the origins of the universe and our planet.

In the beginning, something caused the universe to begin from nothing. How this happened is still a mystery, one that we might crack someday. But what folks like Tyson and Goldsmith know is that at the start of time, everything in existence in the universe was packed into a single point much, much smaller than the head of a pin. Even smaller than an atom. This point of everything packed together was infinitely dense and unimaginably hot. This isn't as crazy as it sounds, especially when you consider what heat and pressure are capable of and when taking into account the fact that matter is comprised of mostly empty space—the emptiness between the nucleus of an atom and the electrons orbiting it.

Allow me to go off on a tangent here to better explain the empty space bit.

A common way to think of atoms in scale is to picture the inside of Madison Square Garden, the basketball and hockey arena in New York City. Place a basketball at the center of the court, smack dab in the middle of the arena, and call it "the nucleus," the collection of protons and neutrons that make up an atom's center. Next, take a golf ball and place it in the highest bleacher seat farthest from center court as you can get. The loftiest of the nosebleed seats. The golf ball you shall dub "the electron." And there you have a rough-scale model of an atom, with all the emptiness lying between the basketball and golf ball representative of the emptiness separating an atom's center from its ring of electrons.

If it were possible to eliminate all this empty space, you could manage to compress an atom to a much smaller area than it occupies now, right? Now imagine if you could do this for all the atoms of a whole solid object. What would that be like? Well, according to physicist Brian Greene, doing this very thing to the Empire State Building (just blocks away from your model atom inside Madison Square Garden) would enable you to compress that entire skyscraper down to the size of a dime.[2] It would be enormously heavy, though, weighing the same as the building weighs now.

For another visual, imagine what you could accomplish with simple heat and pressure. Could you fit the water from an Olympic-size swimming pool into a thermos? I think you can, but not while it's liquid. I'd first superheat it, flash all the water in the pool to steam. Then, I could imagine forcing all that gas—water vapor—into an ultra-strong thermos powerful enough to withstand both the enormous heat and the immense pressure of the flash-steamed pool. I doubt any sporting goods stores or coffee shops sell such an accessory, but if one did exist, then I'm confident you could eventually cram that entire swimming pool into something small enough that you could conceivably carry in the palm of your hand, assuming you were strong enough to hold all that water weight in one arm.

So back to that single point that Tyson and Goldsmith describe as the starting microscopic pinpoint that's the beginning of the universe.

The whole sum of everything in existence was once packed together in a similar sense, only much more so, because the empty space itself didn't yet exist, and the very components of the atoms themselves hadn't been formed yet. Hopefully it's easier now to visualize how heat and pressure could also help to describe the universe's earliest state, except the pressure involved was much, much more crushing. And the temperature of all this? Well, according to the experts it was something on the order of 10,000,000,000,000,000,000,000,000,000,000,000,000,000,000,000,000 degrees Celsius. This is somewhat close to describing the state of affairs that the universe first found itself in during its deep and distant past, approximately 0.0000000000000000000000000000000000000000001 of a second after the Big Bang got started. But get started it did.

What happened next was the hot and dense universe's rapid inflation. Scientists like to use the word "inflation" because they don't want us thinking in terms of an immense explosion with debris flying every which way. By inflation, they mean the universe expanded at a rate much faster than the speed of light, possible because it was space itself doing the expanding, or inflating (the speed of light is only the speed limit of travel within space; space itself isn't bound by the same rule).

It's during the early era of inflation that the four physical forces emerged to begin taking hold of the universe's fate. First, Tyson and Goldsmith explain, gravity appears, becoming the first force to declare independence from the universal unified force that was the nature of the universe at its earliest starting point. Then, two other physical forces broke free from one another, the strong nuclear force (what binds atoms and their components together) and something called the electroweak force (whatever that is—go ask the experts). At this point the universe was remarkably simple: incredibly hot but smooth and almost perfectly uniform. But not 100 percent so.

While the universe was expanding (inflating), it was cooling at the same time. Now think: immense energy source starts to cool. Energy cooling. Matter is energy, cooled (frozen even). Thus, it's during this next phase in the evolution of the universe that matter began to form, only it wasn't quite that simple. The super-hot photon energy comprising the universe began forming matter as it cooled but could do so only in pairs: matter and antimatter. As Dan Brown taught us in his novel *Angels & Demons*, matter and antimatter cancel one another out, annihilating on contact and leaving nothing behind but a burst of energy powerful enough to destroy the Vatican. Except for some reason, our super-hot, rapidly expanding, ultra-young universe wasn't uniform or symmetrical enough to ensure that a perfectly even number of matter and antimatter particles were formed in this soup. It was close, but for every one billion antimatter particles formed, another one billion and one matter particles were created. A very slightly lopsided ratio. So after the matter-antimatter chaos finished self-annihilating, all that was left was that extra bit of matter, the same matter that you and I and everything else are made up of. It sounds kind of crazy, I know, but Goldsmith and Tyson (qualified and experty and all as they are) and many other talented astrophysicists and other smart folks like them assure us that this has all been confirmed in laboratory conditions.

Then the baby universe further expanded and cooled until the electroweak force split into electromagnetism (electricity and magnetic fields, obviously) and the weak nuclear force (responsible for nuclear decay). Meanwhile, matter began cooling and coalescing into the elementary subatomic particles some of us remember from our chemistry classes: protons and neutrons, with electrons free-flying all about in what Goldsmith and Tyson describe as "an opaque soup of matter and energy."[3] Thus, all the basic ingredients were in place: the four physical forces, photons, electrons, protons, and neutrons (clumping together as it all cooled down further). How long did all of this take? About three minutes, we're told. Seriously. Or, as the humorist and author Bill Bryson cleverly put it in his book *A Short History of Nearly Everything*, "in about the time it takes to make a sandwich."[4]

After more cooling and expanding, eventually the electrons slowed down enough to get captured by atomic nuclei, forming the first atoms and the lightest elements on the periodic table: hydrogen, helium, and lithium. It was at this point that our early universe became, Tyson and Goldsmith say, "transparent to visible light," or the photon energy waves that we see when empty space is disturbed or excited. Thus "let there be light," so to speak. This earliest light is still around with us today, only it has grown so faint that it's no longer actually light that's visible to our naked eyes. Rather, these photon waves have dissipated over the billions of years that have passed since they first appeared, to the point that they are detectable only in the microwave wavelength. Scientists have dubbed this still-existing evidence of the Big Bang as cosmic microwave background radiation.

From here, it was off to the races, largely thanks to our old friend gravity. If you think about it, gravity is the ultimate source of energy that makes life possible, since it causes matter to clump together, generating pressure and heat, the same pressure and heat that powers stars and is responsible for the geothermal heat that powers hydrothermal vents.

Thanks to gravity, the earliest elements clumped together to become small stars (which are actually massive), then massive stars (more massive still), and also black holes. And then, all of this clumped together a bit more to form the galaxies.

Stars, as I hope you may know, avoid collapsing in on themselves thanks to the energy generated from nuclear fusion at their cores, but they can keep this up for only so long. Eventually, they start burning out their fuel, gradually dying in steps that fuse the lightest elements stars are made of (hydrogen, helium, and lithium) into heavier elements as the stars get hotter and hotter. Eventually, they run out of this burnable stuff altogether, the center no longer holds, and some of these stars then explode in the most violent and spectacular event in the known universe: a supernova. Supernovae are extremely violent and catastrophic, hot and powerful enough to produce even more heavy natural elements that eventually make life as we know it possible.

Star explosions, supernovae, permitted the universe to become polluted with all those other heavy elements formed during the stars' death throes. That stuff got scattered all over the place. Then, gravity took hold yet again, only next it clumped all this other matter together to form more than just stars, but all the other flotsam and jetsam that orbit those stars today: comets, asteroids, and most importantly, planets.

And all sorts of planets are formed from this very natural process that takes billions of years to unfold, creating a wide variety of worlds so numerous that they are probably uncountable. Some are very big; some are relatively small. Some are very hot, and some icy cold. And occasionally, worlds like ours are formed as well, the so-called Goldilocks planets that seem "just right," at least to the organisms that evolved on them.

In essence, from an immensely dense, immensely hot, near perfectly uniform state at its foundation—wrested from nothingness, somehow—the universe has been expanding, very slowly cooling, and falling apart ever since. But because the force of gravity was created in this calamity as well, little islands of heat and energy are formed in the duration, throughout this excruciatingly slow demise of the universe. And even these heat islands don't last forever; they eventually fall apart themselves, thus conforming to the law of entropy.

Gravity is critical during this entire process; it is the only reason why stars and the planets they give birth to can exist in the first place. The universe was created, then cooled, and then condensed very slowly. Gravity took hold, and the first stars were created then burned out and collapsed or exploded under the laws of physics. Then the second generation of stars formed, together with the planets and other stuff created in the bowels of the first-generation stars as they approached death. They, too, eventually fall apart; second-generation stars collapsed or exploded, or even fizzled out lamely, consuming or discarding the planets and other objects that orbit them in the process. And on and on. Endlessly churning, over and over again.

But before it all falls apart completely, life formed on the surface of a particular Goldilocks planet in some unassuming, unimportant corner of the universe. Life—another type of heat island that gravity makes possible.

So that's the gist of Tyson and Goldsmith's brief introduction to the birth and evolution of the universe. At any rate, from this multibillion-year process we know of at least one planet where life has emerged: our own.

How it happened precisely is the subject of numerous theories, but in all honesty, no one can really say for sure. And we may never know because the Earth is literally churning itself inside out, with entire continental plates submerging beneath others, sinking into the Earth's mantle as they get melted and crushed into oblivion. It's entirely possible that the one fossil that could have enabled us to determine how life got its start here has met this same fate long, long ago. We simply can't say one way or another.

But what is known is that the essential ingredients for life (water, carbon, some other stuff) are abundant not only on this planet, but in our solar system as well. And that means these ingredients are abundant throughout the universe too. How about the so-called building blocks of life, amino acids? They've been discovered in meteorites, rocks that have been orbiting the sun for what must have seemed like an eternity until the day Earth's gravity captured them and pulled them to its surface. Thus it's a safe bet that amino acids are abundant in the universe as well. And though we can't say for certain how this recipe for life came together and successfully cooked up the first living organism on Earth, a process by which this could have occurred has been replicated in the laboratory environment with great success. Yup, life of some sort has been "created" in the lab by eminent and scholarly individuals on a quest to work out how it came to pass for this planet and thus for us. It's not proof of precisely how it happened in the ancient past, but it's an interesting proof of concept nevertheless.

Once life springs forth, the forces of evolution by natural selection take hold.

Evolution is a relatively simple process. It's driven entirely by chance, by probability.

Because of something called entropy—the tendency of matter and energy to move from a state of order to disorder—living things eventually weaken and die. Living organisms on Earth's surface are basically a means developed by organic chemistry to capture, store, and distribute the sun's energy for a little while. But because each single living organism won't live forever, reproduction emerged to allow life to go on and store this solar energy for a bit longer. This is where probability comes in.

Imagine you have a computer file, that short story you wrote but never published. Copy and paste it to your desktop. Now make a copy of that copy. Then make a copy of that copy of a copy. Next make a copy of that copy of a copy of a copy. Keep this up for 100 generations, maybe even 1,000 or 10,000, and perhaps you'll find that the newest copy is an identical match to your original file. But this cannot last forever—perfect fidelity won't hold. Eventually, after enough generations of this, you'll notice errors setting in, making the newest generations slightly different from the older ones. Maybe the protagonist's name is spelled wrong in one place. Or a paragraph isn't indented right. This happens because it *can* happen, because there's a probability that an error will result from this process, even if that probability is slight.

The sort of life that gave rise to our kind comes about through sexual reproduction—two files combine their information to create a new one that shares traits of both parents. I've inherited my mom's blonde hair and blue eyes and my dad's eyebrow ridge and jaw line. When sexual reproduction enters the picture, there's an even greater probability that errors will occur in the process of life replicating itself to continue its existence. These chance errors, or mutations, occurring in a constantly changing environment are the force driving evolution. Errors or mutations occur because probability dictates that they can occur, even if they are extremely rare—given enough time, the improbable becomes the inevitable.

Errors or mutations come in three flavors, determined by the environment they occur in. Some mutations are neutral, meaning they don't affect that organism's chance of survival and reproduction one way or the other. They're relatively minor, insignificant changes that go unnoticed by the forces that determine survival. Other mutations are negative, meaning they make it less likely that an organism will survive long enough to reproduce and pass on these negative traits to the next generation. Because negative errors lessen a creature's chances of survival, whatever their environment, those individuals tend to die out before having an opportunity to reproduce, thus these negative traits die out with them (or tend to).

Then there are the positive errors, meaning changes or mutations that are neither negative nor neutral but actually enhance an individual organism's chances of survival in a habitat or environment. Because these mutations end up improving fitness and survivability, the individual creature holding these traits has a better chance of survival and successful reproduction, meaning the odds favor these positive traits being passed on to the next generation and beyond.

Let's consider a stupid hypothetical example: hairy chests in humans.

Suppose there is a solitary group of women and men living a simple life on Earth in a primitive society, a venerable Garden of Eden, only much more crowded than the Biblical version. In this society, all the men have hairy chests. That's just the way it is, and the women are accustomed to this and comfortably deal with it. And life goes on for several hundred years or so like this, with each successive generation of men adorned with glorious chest hair, and the women of those generations accepting this trait and bearing children with these men, as humans are prone to do. That is, until one day, for reasons totally unknown, one boy becomes a man but does not grow any hair on his chest. No, this man's chest is entirely uncarpeted. An error occurred in the process of sexual reproduction.

His buddies rib him about this, of course, but as it turns out, the ladies absolutely love this new look. In fact, it drives them completely mad. And this of course translates to . . . well, let's just say our smooth-chested friend is never lonely on Friday nights.

He sires several children, and his male offspring share the trait of their father, so when they grow up, they too have no chest hair, as their father didn't, increasing the number of hairless-chested men in the population at large.

What next? Well, these boys too are more successful in love—the females in this society are supremely attracted to this new hairless look and are unmoved by the posers who fake it and try to hide their hairy ways by shaving (the stubble and sandpaper feel are dead giveaways). The naturally hairless sons find willing partners, becoming fathers to more children, and the men of that next generation also have hairless chests. And on and on and on. Yes, the hairy-chested men still father sons of their own but not nearly as many, and those poor boys have difficulty competing for the affections of the females in the group due to the rising population of hairless-chested men.

Eventually and over enough time, the hairless chest wins over the hairy in the war of romance, and after a few generations, most of the individual men in this society have hairless chests, outnumbering the hairy chested. Things go on like this until eventually hairless chests are the defining feature of the males in this society, completely opposite of what it was generations prior. Ultimately, only one hairy-chested man remains, a freak of nature who dies a lonely bachelor, childless.

That's essentially how evolution is driven. It's a simplified, stupid way to describe it, so my apologies, but I think it works. There are probably millions of individual traits that could stand out, expand, and eventually dominate a population. I arbitrarily chose chest hair (or lack thereof) as an example.

And the driving force favoring the emergence of any trait could be any number of things, not just changes in sexual preference. It could be a change in diet or a change in environment or climate. It could be external pressure from another population of animals or separate populations of the same species clashing. Parasites or diseases could even come into play as a force for evolutionary change. Whatever the precise process, it's an extraordinarily simple premise: traits that are beneficial to a population's survival or that engender better reproductive success for an individual assert themselves and have better odds of being passed down to the next generation, translating into a countless number of subtle changes to living creatures (animate organic chemistry, essentially) and eventually adding up to the amazing variety and diversity of life on Earth today.

With almost four billion years of this going on, one easily could imagine how this slow, relentless, blind but nonrandom process could take life in the form of single-celled bacteria, whether from a hydrothermal vent or elsewhere, and gradually generate from that tiny aquatic worms, then fish, then amphibians, then reptiles, then mammals, to monkeys, to apes, and to us: Homo sapiens, the world's only surviving species of a bipedal, mostly hairless, talking ape that occasionally demonstrates impressive levels of intelligence. Blind, unplanned evolution by natural selection permits organic life to gradually morph from one species to a different species, and on and on in an unbroken chain. That's where our planet's fantastic biodiversity comes from.

All the above—a 14-billion-year process—gave rise to the dodo, a species that humans wiped out of existence in approximately 90 years. It gave rise to the vaquita, the world's smallest porpoise, a species that may not survive beyond this decade. And this process gave rise to us, to our kind.

We're here by chance, the probability of reproduction giving rise to mutations that may prove advantageous to individual and group survival. Every other living organism on our planet arrived here the same way. They are our cousins. Against all odds, life found a way to emerge and thrive on this big blue marble we call home. We're the only creatures on this planet aware of these details. And yet we're failing our cousins, our cohorts in the animal kingdom and the photosynthesizing plants at the very foundation of the surface food chain. We can do better, for sure. We're different, unique. Ours is the only species that mourns the passing of other species. We have drafted local and national laws and even international treaties in an effort to halt the human-driven extinction crisis. We've succeeded in many areas, but on the whole, we are unfortunately failing. Now we are systematically destroying Earth's biodiversity. What took eons to produce, billions of years, we're deliberately annihilating in a generation or two.

"Giraffes, parrots, and oak trees, among many species facing extinction," reads an August 7, 2022, item from *UN News*. Citing the latest assessment data by the International Union for Conservation of Nature (IUCN), the report shares such dismal news as the fact that more than 30 percent of the world's known species of oak trees are now facing extinction, mainly due to logging. Another 41 percent are of "conservation concern," according to IUCN, as a rising human population fells more trees to make way for crops and settlements.

I easily spotted giraffes during my travels in Kenya and Amboseli. Spotting a wild grazing giraffe is a remarkable sight to see, and I'm truly grateful that I got to witness it with my own eyes. Apparently, I'd have much less luck were I traveling in West Africa—IUCN estimates that there may be fewer than 600 West African giraffes roaming wild, not 600 species but 600 individuals of a single species. A growing human population in those countries hunts them for food. The ones that aren't shot are losing habitat as more trees are felled for agriculture. It's the same sorry excuse—extinctions driven by humans consuming habitat, leaving little for wildlife.

The *UN News* item I reference here was prompted by the release of that periodic ecosystems assessment report published by the Intergovernmental

Science-Policy Platform on Biodiversity and Ecosystem Services, IPBES. For sure, the UN's internal news division followed the release. Few other outlets did. The report itself emphasizes the risks to humans from species loss and overexploitation and advocates for sustainable harvesting so that impoverished populations can continue enjoying the economic benefits to be derived from the exploitation of species. But can these vulnerable species ever be sustainably harvested in a world that now harbors eight billion humans, the species responsible for this biodiversity loss disaster in the first place?

The IUCN Red List categorizes species one of eight ways: (1) extinct (the Yangtze River dolphin or the Guam flying fox, both declared extinct in 2019), (2) extinct in the wild (as the California condor was once, or as the Hawaiian crow is today, and the Tasmanian devil may soon become), (3) critically endangered (the vaquita of the Sea of Cortez), (4) endangered (like the Grand Cayman blue iguana), (5) vulnerable (bull sharks), (6) near threatened (like the rainbow parrotfish, a species that's critical to coral reef health, which I've spotted only in conservation areas while scuba diving), (7) least concern (the Japanese serow, an amazing recovery story in its own right), and something referred to as (8) data deficient, meaning science simply doesn't have enough data on these species to make a firm determination as to their status (wild cabbage). I was once optimistic about that eighth and final species category, but I'm not anymore. Recently, critics have argued that the eighth category shows that there's something off with the way IUCN and its partners conduct Red List assessments. Basically, they argue that existing methodologies are too analog in approach. It may be time to utilize the best technology available to get a firmer grasp on how badly humanity is killing off other creatures.

A new study by a team of researchers in Norway predicts that nearly half of data-deficient species are actually threatened with extinction and thus nothing to be optimistic about. I suppose that would explain why there isn't enough data on them. They say their advanced tech-driven modeling shows that 85 percent of data-deficient amphibians actually may be facing extinction now. The research team behind this new study argues that their findings suggest IUCN needs to change its listing methodology, perhaps by incorporating the use of machine learning technology that factors probability back into the equation, the very thing that makes biodiversity possible to begin with.[5] In essence, they say the usual tools IUCN uses to assess 140,000 of the millions of species in existence—known range, known occurrences, population size estimates, and population trends—are simply not up to the task, arguing that it's time to rely on some more advanced technologies, especially the powerful computing tools and computing models developed in recent years. "The sheer amount of known and unknown species globally, the dynamic nature of threats and trends, and limited human resources for undertaking such Red List assessments turn this critical endeavor into a Sisyphean task," they said.[6]

Will technology eventually save the day? Many are hopeful that it will. In fact, some believe that you and I may very soon see the dawn of de-extinction technology.

A team at the University of Melbourne and Colossal Biosciences, a company based in Dallas, are collaborating to revive the thylacine, more famously known to you and me as the Tasmanian tiger. A species that was once wiped off the face of the Earth may soon walk its surface again, à la Jurassic Park. You remember the thylacine. I mean, you don't, because they're extinct, but you've almost certainly heard of this animal. Wait around awhile, and you may be able to actually go see a living one.

This species thrived on the Australian mainland for thousands of years until something—almost certainly the first human inhabitants of that continent—caused them to become entirely extirpated from that land mass roughly 2,000 years ago. Tasmanian tigers nevertheless survived on the island of Tasmania in a small population until they were rendered extinct from the wild almost 100 years ago. Hence the "Tasmanian" in the name, though their natural habitat originally wasn't limited exclusively to that island. "Tiger" came from the famous stripes that ran down their backs, even though they were in fact marsupials more similar in appearance to canine species. This fact alone tells us that the European settlers responsible for the thylacine's extinction weren't exactly the sharpest tacks on the wall. Imagine if someone had first laid eyes on a male lion in the African savanna and declared it the "Tanzanian poodle."

According to the website Live Science, in an August 24, 2022, article, University of Melbourne scientists say they've sequenced the entire Tasmanian tiger genome and have enough viable DNA to revive the species with the right technology. Their partner, the genetic engineering company, now says that technology is within grasp. Together, they think they can give birth to a baby thylacine using a surrogate mother sometime in the next few years. The collaboration envisions Tasmanian tigers returned to the wild within a decade.

How serious are they? The Melbourne lab calls itself the "Thylacine Integrated Genomic Restoration Research Lab," or TIGRR Lab. The news report said they already have identified a closely related species, the fat-tailed dunnart, to use as a "template genome" from which they can re-create a viable thylacine embryo.

The Tasmanian tiger could soon live again. But should it? This gets me back to the Jurassic Park analogy—the scientists involved in this endeavor certainly seem confident that they can. Have they seriously considered whether they should? It's a valid question, especially considering the ongoing extinction crisis and the thousands of species on the brink of extinction today. Perhaps resources are better spent protecting these creatures—on preventing extinctions—rather than simply reversing extinction. And is the habitat available on Tasmania still suitable for the species to thrive and reproduce there? Is it sufficient to maintain a reasonable self-sustaining population size? Will they be returned to the Australian mainland? After all, people probably destroyed them there, too. Or are we bringing these creatures back from oblivion just to stick them in zoos, ostensibly for their protection but more likely for our amusement? Is de-extinction technology being pursued to right thousands, perhaps millions, of wrongs? Or are there other motivations?

This self-doubt and navel-gazing could go on endlessly. Perhaps there are other, more deserving extinct species to revive from the great beyond, species that could fill a needed ecological niche, for instance, or that could help beat back a particularly nasty invasive species problem. I sometimes muse that perhaps invasive species problems could be tackled by deliberately introducing another invasive—say, find out what keeps the lionfish under control in the Pacific and simply relocate some of that to the Atlantic. But with de-extinction technology, perhaps this isn't necessary—maybe the Caribbean monk seal would've ended the lionfish infestation easily were it still around. Could that species be revived? Should it?

On the question of the Tasmanian tiger, I say do it. I say this mostly for selfish reasons, of course. I've seen photos and even video footage of this species, and I want to live in a world where I know this beautiful animal is alive and well. If the Tasmanian tiger were ugly, would I say the same thing? Is there a scientific team out there soldiering away earnestly to revive some of the world's least-easy-on-the-eyes endangered species? Well, for now, who cares? If we have the technology to thwart our ancestors' carelessness, callousness, and stupidity, then we should employ it toward these ends, I believe. We can't correct all the mistakes of the past, but we can choose a select few to reverse. So long as we're honest about our motivations. And after we've carefully considered the potential consequences. Let's not bring back velociraptors, for sure. Passenger pigeons, great auks, and dodos? You know, I can't see the harm, but someone far more qualified than I may have objections, and these should be seriously considered.

Still, the very concept of de-extinction is extremely enticing, one of the things that made the original *Jurassic Park* novel and movie so popular until

Hollywood got carried away and made something like a dozen spinoff movies (seriously, folks, stop going to Jurassic Park—people get eaten there!). But if the technology does exist, this might be a good thing, in balance, perhaps the only real way to ultimately save the vaquita.

Our kind and the tremendous biodiversity that exists today arrived in this universe by chance—energy created fantastic order while defying entropy, all through a sequence of chance occurrences. Extinction is inevitable. For thousands of years, humans have been actively helping along the entropic forces leading toward extinctions. Perhaps the time finally has come for our kind to do all we can to help our animal cousins beat back extinction, if only for a little longer. Chance brought all species here, our kind included, but now we may have an opportunity to tip the odds toward life's favor a bit.

Woolly mammoths? Saber-toothed tigers? Giant sloths? That car-sized armadillo thing? Are these candidates for de-extinction? Hmmmm . . . let me think about it.

Where do our fine-feathered friends *tancho* and whooper stand today? They're doing just fine, though they certainly could be doing better.

It's long been a dirty little secret that America's wildlife is laced with organic and inorganic pollutants and heavy metals, a consequence of our air pollution finding its way into the waters, vegetables, and animals that they consume. In some places, it's caused by pollution from mining. Sadly, *tancho* hasn't escaped this fate either. Pollution concentrations in red-crowned cranes rose along with Japan's industrial rise. Concentrations dropped as Japan's air and water problems improved, but the issue still persists. A 2015 study discovered a slow but steady accumulation of mercury in red-crowned crane kidneys. It's a discouraging setback for sure. "Previously, we reported that red-crowned cranes in Hokkaido were highly contaminated with mercury in the 1990s and that the contamination rapidly decreased to a moderate level in the 2000s," the researchers recalled.[7] Their study doesn't definitively pinpoint where the mercury is coming from but merely speculates that it might have something to do with historic mercury mining operations that previously existed in parts of the region that *tancho* calls home.

Mercury is hardly the only threat. In 2018, researchers discovered that red-crowned cranes are increasingly contaminated with persistent organic pollutants (POPs), especially polychlorinated biphenyls (PCBs). "Detected POP concentrations were at levels that negatively affected health in other avian species," the researchers reported. On a positive note, they say the POP concentrations were found to be lower than some species, probably because red-crowned cranes are omnivorous, whereas bird species suffering higher POP concentrations in their tissues tend to be strictly carnivorous.[8]

Mercury may be a greater threat to the whooper.

One thing left unsaid in this entire discussion is that, aside from its massively long migration undertaken twice a year, parts of the AWB whooping crane's stomping grounds are in the oil patch. I mentioned oil wells located right inside the Aransas National Wildlife Refuge. These are old marginal wells, however, probably pumping only a few barrels a day by now. But when these birds have had their fill of blue crabs in Texas, the whooping cranes make their way through Oklahoma, then parts of Kansas, where some drilling activity happens. They're not known for populating any portion of northeastern Colorado, which has enjoyed an oil boom in recent years, but as their numbers grow, a few whoopers probably will start turning up there, as well. They're fond of the Platte River's mudflats in Nebraska. From there, they likely follow much of the Missouri River's run, which would take them in the vicinity of Williston, North Dakota, the capital of North Dakota's modern shale oil boom (which should be losing steam right about now, if U.S. federal government predictions prove accurate). And, of course, this fair-weather bird spends a significant chunk of its summer in northern Alberta and Wood Buffalo National Park. The park is famous for being Canada's largest in total area. It's also downstream from the Athabasca tar sands, the center of Alberta and Canada's oil sands industry.

Pulling crude from tar sands, or oil sands as they're more popularly known, isn't like hydraulic fracturing or conventional oil operations. In much of the region, the oil sands aren't drilled but mined. The bitumen-laced sand is dug from the ground and then hauled off to processing centers where it's essentially cooked, releasing the oil. Another method is called "steam assisted gravity drainage," or SAGD. Here, water is boiled to produce steam, which is then injected into the sands underground at high pressure. The oil is released and then percolates down, where the industry has inserted special collectors to capture and move it. Both methods involve a lot of energy, especially natural gas burned to generate heat and steam. Both methods release tons of carbon dioxide into the atmosphere, aggravating global warming. And both methods release considerable amounts of mercury. The Athabasca River flows through an active patch of oil sands surface mining activity, collecting mercury along the way, which it then carries to Wood Buffalo National Park as it flows northward.

I found nothing in the scientific literature that suggests whooping crane reproduction is being inhibited by the presence of environmental pollutants in their tissues. But it's an issue that warrants monitoring in the future, especially if it gets worse. A Canadian study published a few years ago confirmed a probable link between Athabasca tar sands mining and processing operations and mercury levels in whooping cranes, though the authors of the study mentioned below cautioned that no definitive link or firm proof had been detected. But that's the most likely source.

All we know is that whooping cranes are becoming contaminated by elevated levels of mercury during their time spent in Canada. How? By measuring the mercury content detectable in whooping crane eggs. During periods of heavier rainfall and higher water flows, mercury concentrations rose—not only for whooping cranes, but for a host of aquatic birds that rely on the same prey the whoopers do. "For the majority of species/site analyses, higher egg [total mercury] concentrations were observed in eggs laid following years of high river flow," wrote Craig Hebert of the National Wildlife Research Centre in Ottawa. "Hence, flow was likely of critical importance in moving contaminated sediments to downstream areas with resultant impacts on the bioavailability of Hg to wildlife such as birds."[9] Hg is mercury. Again, the caveat to avoid the inevitable lawsuit: "This study provides no direct evidence linking Hg levels in eggs to oil sands sources. However, previous studies have demonstrated that oil sands developments are a source of Hg to the local environment." For example, we know that more than five times the amount of mercury can be detected in snow within a 50-kilometer radius of oil sands operations than outside that radius, and water flowing through oil sands developments in summer contains triple the amount of mercury than in other areas.[10]

Hunting is still an ongoing concern in whooping crane rehabilitation, much less so for the red-crowned crane in Japan. Hunting is allowed in Hokkaido and even encouraged. Because the Japanese (with American assistance) destroyed the Hokkaido wolf, there are few apex predators left on the island to take care of the Ezo sika deer, one of the world's largest deer species. I've had several up-close encounters with Hokkaido's deer, which to an untrained North American eye easily could be mistaken for an elk given their size. Once, my wife and I were eating at a favorite ramen joint on the east side of the city across the street from a large defunct coal mining operation, when a herd of deer suddenly popped into view, dashing across the street and startling cars and pedestrians alike. I joked that it looked like Santa's reindeer had escaped and were running loose.

The island is believed to be overpopulated with deer, so authorities welcome hunters to thin the herds a bit. Local restaurants increasingly offer Hokkaido venison on their menus in a bid to encourage more trade. But guns are difficult to get in Japan and hard to hold on to. You generally need to belong to a hunting club, and it can take months or even a year to clear the paperwork. Even then, the police check in every year to ensure that you're properly storing the weapon and all your paperwork is still in order. Mess around, and you forfeit your privilege to own a firearm in Japan. We heard about a frustrated and possibly drunk farmer in the Tokachi region shooting a red-crowned crane dead because it wouldn't leave his cattle feed alone. I was quite surprised—guns are rare, and red-crowned crane poaching is effectively nonexistent. Upon further investigation, it was revealed that the attack was conducted with an air rifle, not an actual gun.[11]

Contrast this to the United States, where all you need to obtain a high-powered semiautomatic weapon is a library card. I'm exaggerating here, of course, but not by much. Studies have found that hunting has seriously impacted efforts to revive whooping crane populations in the east. For instance, a small population in Florida used to migrate to Wisconsin, aided by a specialist guiding them in an ultralight aircraft (last I checked, this expensive program was discontinued, and the population was left to reside permanently in Florida). Hunting likely has hit that flock badly. We also know that hunting has been a thorn in the side of the folks trying to increase whooping crane numbers in Louisiana. But there is no indication that hunting is a contributing factor to the relatively lower reproductive success rate in the AWB whooping crane flock compared to Japan's red-crowned cranes. Of the monitored whooper flocks in North America, experts agree that the AWB whooping cranes have been least impacted by careless people with guns than other populations.[12]

But with sandhill crane numbers now easily in the hundreds of thousands of individual birds, the pressure to open this species to hunting is mounting. Sandhill cranes (and whooping cranes) were initially afforded protections under the 1918 Migratory Bird Treaty Act. Some subspecies of sandhill cranes are still protected under the Endangered Species Act. But sandhill crane hunts are authorized to some degree in at least 15 states and probably more. The number of states where you can hunt sandhill cranes is likely to increase. Humans, as usual, prioritize our needs over the rights of other species, and the sandhill cranes are increasingly being blamed for damaging farmers' crops. By some estimates, a single sandhill crane can consume some 800 planted corn seedlings in a day, and 100 sandhill cranes could theoretically wipe out eight acres of crops over a weekend.[13] Some level of hunting can almost certainly be sustained and may even be called for in some circumstances. The trouble is that in the past, hunters who killed whooping cranes pled innocent, insisting that they thought they were shooting sandhill cranes instead. Some even tried to say they thought they were shooting at "albino" sandhill cranes, which is nonsense, of course, but some judge in Oklahoma or Wisconsin may have a more forgiving take than a person like me.

Climate change may be a greater long-term threat to the whooping crane, less so for the red-crowned crane.

The changes caused by global warming almost certainly include erratic weather to eastern Hokkaido. I've witnessed firsthand bizarre and intense rain episodes that clearly caught the locals by surprise, flooding the swamps and streets alike. This shouldn't be too surprising as most of Kushiro is low lying, since it was built atop a wetland. But the region is more accustomed to seeing its precipitation arrive in the form of snowfall, not drenching rains. Hokkaido as a whole is famous for escaping the annual rainy season that hits the rest of the archipelago. Some suggest this could be changing now, at least for the southernmost portions of the islands. Warmer Pacific Ocean temperatures may make it easier for typhoons to reach Kushiro, just as Atlantic Ocean hurricanes are starting to reach as far north as Halifax, Nova Scotia. But all in all, folks in East Hokkaido seem to like the fact that the winters appear to be getting shorter and less mean. *Tancho* probably likes this, as well. Shorter winters should bring more abundant forage to the region's wetlands, suggesting that climate change could prove a boon to the red-crowned crane over time. But climate change poses a potentially serious risk for AWB whooping cranes because of the seasonal migration they endure to facilitate their survival.

"Warming temperatures associated with climate change can have indirect effects on migratory birds that rely on seasonably available food resources and habitats that vary across spatial and temporal scales." So warns a research team from the U.S. Geological Survey and the University of Arizona. Relying on two indexes designed to gauge when spring starts, the "first leaf index" and "first bloom index" or FLI and FBI, they sought to estimate how earlier onsets of spring attributed to climate change could be impacting important national wildlife refuges and the flyways for the whooping crane and blue-winged warbler, another migratory avian species. Not surprisingly, they detected big changes occurring over time, especially in the higher latitudes. "Our results show that relative to the historical range of variability, the onset of spring is now earlier in 76% of all wildlife refuges and extremely early (i.e., exceeding 95% of historical conditions) in 49% of refuges," they discovered.[14]

It's a serious risk.

I noted earlier that by far the greatest factor driving our modern-day extinction crisis is the loss of habitat available for wild animals and plants to thrive in. We humans have consumed so much of it, and the patches that we haven't completely destroyed (lands that are either deliberately set aside for conservation or areas that are too high, too dry, or too cold for our liking) are isolated from one another in a busted-up quilt work, an "islandization" of the continents. Islands are generally poor laboratories for evolutionary biodiversity. Now climate change has entered the picture, threatening to shift climate zones and weather regimes in different directions. Hot and dry regions may become hotter and drier, while wet regions could become even wetter. Droughts are more pronounced and last longer, interrupted by catastrophic flooding.

The Aransas National Wildlife Refuge may be nice from December through April today, at least as far as the whooping cranes are concerned, but will these conditions last? Will their seasonal forage be plentiful enough? Will there be enough blue crabs left when they arrive, or will these all have been eaten by year-round resident species? And what should happen if the whooping cranes turn up at Wood Buffalo National Park too late for the start of spring? Will this affect their breeding and overall reproductive success? These researchers fear that it might, given the vast journey the whoopers undertake twice a year. "The metabolic demands of excessive flight journeys mean that migratory and reproductive success hinge on the availability of sufficient food resources and optimal habitat conditions at stopover locations and upon arrival at breeding grounds," this research team points out. "It is unclear whether shifts in migratory timing are able to keep pace with alterations in plant phenology and food resource availability across broadly distributed habitats."[15]

Of course, artificial feeding could help mitigate these changes to some degree, but here I digress. Besides, food availability is just one consideration that wildlife refuge and parks managers need to juggle as ever-increasing atmospheric concentrations of greenhouse gases deliver steadily rising temperatures, as the laws of physics demand. Invasive species problems could get worse. Diseases may propagate more easily—perhaps that parasite problem becomes more acute, more deadly. Other wildlife may move in and out at different times and in different locations, upsetting some natural balance whoopers have grown accustomed to. An even greater impact may result from the human presence: warmer weather could equate to ever greater numbers of visitors enjoying Aransas and Wood Buffalo, not necessarily doing the antisocial crane any favors.

Climate change threatens to alter local climates and even entire ecosystems. These changes may be too rapid for species to cope, even highly migratory, omnivorous species like the whooping crane.

I'm not too worried, though.

Thanks to the Endangered Species Act, the whooping crane will endure, thriving in some corners of North America in self-sustaining population numbers, perhaps expanding well beyond their current limited ranges as their numbers swell, as with the red-crowned crane of East Hokkaido. My young nieces will get to enjoy them long after I'm dead. One day, they'll probably be removed from the official list of endangered species, perhaps not long from now. It's too early to do so today, in my opinion. But I don't get to make these decisions. By 2035 the AWB whooping crane population probably will be in the range of 1,000 individual birds, if not more. The Florida population probably will no longer be with us, given that neither the U.S. federal government nor the State of Florida seem all that inclined to spend money to protect them (as far as I know). But there are careful stewards of the flocks in Louisiana and Wisconsin. They'll do OK—not as well as the AWB whooping crane, but well enough—so long as officials can convince hunters to keep their rifles aimed squarely away from them.

The last time I saw a whooper in person, I was on my way back to Houston from my grand tour of the Powderhorn Ranch—today the Powderhorn Wildlife Management Area—the newest addition to the cranes' Texas winter home. I pulled a hard right out of the ranch's gates and began accelerating on the county road in the direction of Port Lavaca and the highway that would take me back to the Bayou City. But there was plenty of daylight left, traffic was nonexistent, and the weather was perfect, so my mood put me at the speed limit so that I could soak up the scenery a bit more, seeking excuses to pull over and let the day linger somewhat longer. I soon found one: whooping cranes—two adults and one juvenile—standing in the middle of a field to my right, close to that isolated easternmost and smallest segment of the Aransas refuge. I found a safe spot to pull over and got out of my car.

They were still too far away for a good photograph, especially with the lousy camera I brought with me that day, but they were close enough. They strutted and fed, mindful of my presence but none too concerned given that I was too far away, and they had a very clear line of sight to me, with ample time to escape had I been a predator determined to make a meal out of one of them. A snow-white bird with a striking crimson crown—it's something to behold. The juvenile was brown and mottled, camouflage that helped it blend in as easily with the broken stalks of whatever crop had been harvested there, as with the grasses and reeds of the crane's wetland homes. At one point, all three paused and stared at me from the distance. I stared back, wishing I could get closer but content with the fact that I could soak in this sight without too much trouble when the season was right and it suited me, something deemed nearly impossible back in Hornaday's day.

I shifted my weight to my left leg when I decided to make my way back to the car. This prompted a baby whooper to edge in just a bit closer to its parents. By then, mom and dad whooper couldn't care less about me. Content that I posed no threat, they continued to make their way across the field, pecking at seeds and kernels as they did. "See you guys later," I said, sitting down behind the wheel and slamming my door shut. And it's true—I will see them again, probably before this book publishes. But back then, I knew I was making my way out of Texas and onward to northern Japan, so the whooper and I would be separated for a bit longer than I'd prefer, but that's life.

Tancho will survive and continue to thrive, as well, regardless of the Ministry of the Environment's experiment with artificial feeding.

My last encounter with *tancho* happened on the farm roads just outside Kushiro, close to the national park. They can be seen nearly anywhere in the vicinity—reliably at or near the feeding stations in winter and in farmers' fields or wetter reaches of Kushiro Marsh at any other time of year. The best memories are at the feeding stations. They caw and sing and dance nonstop, a constant flurry of motion. Wait around long enough, and they'll suddenly run toward you, wings flapping, building up speed until they ultimately take flight and then soar above your head, casting their jumbo-jet shadows over the ground as they head west toward open space, away from the crowd and confines of the formal feeding station. They were once almost totally, strictly confined to Kushiro Marsh, the Akan center, the Itoh sanctuary, and Tsurumidai. Now, they're not only spotted at the farms and fields surrounding the national park, but also in the eastern wetlands near Nemuro, in the popular Tokachi area near Obihiro, along the southern coast as far as Tomakomai, and now even in the northwestern edges of Hokkaido at the marshes of Sarobetsu. Someday soon, they will be spotted nesting and foraging in the eastern Aomori and Iwate Prefectures on Honshu. This event will bring an already extremely famous, iconic Japanese species even more fame and notoriety and lend further credence to the view that the Japanese took the right approach to rehabilitating this once near-extinct species, even if they themselves don't necessarily agree with this view.

I'll never get tired of watching the red-crowned cranes dancing in their wintery wonderlands. A snow-white bird, black highlighted wings, with a striking crimson crown—it's something to behold.

To the women and men who took the initiative to save both red-crowned cranes and whooping cranes and who continue to look after them to ensure that they will be around for future generations to marvel at, all I can say is thank you. We owe you an enormous debt of gratitude.

It took nearly 15 billion years for us to get whooper and *tancho*. Thanks to these dedicated women and men, these cranes will be around for many more years to come.

~

The Endangered Species Act is far from perfect, but it isn't a disaster, either. During the 50 years since it became the law of the land in the United States, the ESA has saved scores of remarkable and unique species from oblivion. Any attempts to water down the act should be thwarted, especially given the ongoing pressure a rising U.S. population is bringing to bear on America's last surviving wild patches.

It's important that we dwell a bit on that past point: the pressure of human population.

As with overfishing, ocean plastic, air pollution, soil contamination, and climate change, the rolling global extinction crisis on Earth can be tied to one overarching factor: human population growth. Simply put, there are probably far too many of us, and the more our numbers rise, the worse it gets for wild plants and animals. The United Nations said that in November 2022 the human population on Earth hit eight billion for the first time. A best-case scenario sees it peaking at around nine billion over the next couple decades before ultimately declining to the eight-billion-people range by, oh, the year 2100 or so. More pessimistic outlooks suggest the human population increasing to as much as 10 billion or 11 billion, but I deem this unlikely given recent demographic trends. Still, when people talk about achieving sustainable growth or balance with nature or even when discussing how to halt and reverse the troubling and accelerating decline in biodiversity, rarely is encouraging slower and perhaps even negative human population growth discussed. It may be a taboo subject, but it must be said that part of the reason our planet is in the mess we find it in is because of the massive explosion in human population numbers witnessed in just the past 100 years. A former UN demographer wrote not too long ago that "there is hardly any major problem facing America with a solution that would be easier if the nation's population were larger."[16] Add collapsing biodiversity to that list of major problems.

And yet there are far more Americans today than there were when the Endangered Species Act was signed into law in 1973, and the whooping crane population is higher, not lower. The same goes for other species protected by the act. This is a testament to what's achievable even in a worsening situation—as urban sprawl marches on and as invasive species problems grow worse, many species are nevertheless better off than they were when these situations were less intense, thanks to the ESA and the women and men who implement the law. Yes, there are critics of the ESA. But it seems that for every one of those, there are five or more ESA defenders.

Here's an example. Remember where I wrote about two Canadian researchers who think the ESA has much to learn from Canada's SARA species protection law? They made a few good points comparing and contrasting the two laws. But a few of their colleagues in British Columbia and Nova Scotia take issue with the way they characterize the situation. In their telling, it is the Canadians, not the Americans, with lessons to learn, and the tale of the whooping crane's revitalization under the ESA is one of those lessons to take to heart.

Measuring the ESA and SARA on equal grounds, this research team wanted to compare the laws side by side in terms of both qualitative and quantitative metrics, meaning they wanted to know whether these laws inspired conservation authorities to set certain population size goals, along with other goals deemed important in determining the relative health of endangered species' recovery trajectories. Today the Canadian authorities have a goal of reaching 1,000 whooping cranes summering at Wood Buffalo National Park, but this must be a recent addition to their management strategy—in 2019, Pawluk and colleagues criticized SARA-driven Canadian conservation authorities for lacking ambition in their endangered species recovery goals compared to their U.S. counterparts guided by the ESA. "Comparing between legislation types, ESA recovery goals generally were more quantitative and had higher ambition than those put forward under SARA," they wrote. "While recovery goal quality could be improved under both legislative acts, this appears to be particularly the case for SARA-listed species in Canada."

They go on to acknowledge that the ESA is far older, has withstood an onslaught of legal challenges throughout its history, and generally enjoys greater implementation funding compared to SARA in Canada. Still, their overarching point holds: conservationists in Canada (and elsewhere for that matter) are often too timid and conservative in their strategies and prescriptions for endangered species protections. Bigger thinking and bolder action may be required here, especially considering the awful mess our world is now in. As Pawluk and her research team wrote: "In the midst of a global extinction crisis, effective conservation actions for recovering species at risk are crucial. When setting targets for at-risk species recovery, low-quality recovery goals may undermine or stymie conservation efforts and, thus, are counterproductive. . . . Recovery goals could be improved by focusing on the development of unambiguous, high-ambition (i.e., increasing population), and quantitative recovery goals wherever possible."[17] This article could have inspired the Canadians to adopt their target of 1,000 whooping cranes. And the Endangered Species Act was the original inspiration for this push toward more ambitious recovery goals.

The ESA continues to inspire species conservation in other nations. But as I've tried to make clear, those responsible for implementing the act should pause to consider some of the lessons learned from conservation initiatives elsewhere, especially those cases that have seen far greater success.

Would whooping crane numbers be higher today if the Fish & Wildlife Service hadn't abandoned artificial feeding? I can't help but think that, yes, whooper numbers almost certainly would be greater today if the Americans had taken some inspiration from what was happening in East Hokkaido, even as the Japanese authorities took inspiration from the ESA when drafting their own ACES law. That doesn't necessarily mean that the whooping crane's stewards should adopt the Japanese playbook verbatim, but if the ultimate goal is far more surviving whooping cranes than the current population, one must wonder why this direct population management tool was never revisited. U.S. authorities have spent millions of dollars feeding and breeding whoopers in captivity and then releasing a few of them into the wild, only to see negligible or no impact on overall population sizes. All this time, they could've been spending these resources on a feeding regimen that would effectively serve as a means of improving the reproductive success of the entire population of whooping cranes, benefiting all equally, not just those select few eggs plucked from their nests and chosen for an artificial breeding campaign that ultimately fell to the budget cutter's axe. Japan's *tancho* lovers face no such fears—in fact, the powers that be have been bending to the East Hokkaido locals' will recently, likely because they were the original force behind the remarkable rebirth of the red-crowned crane.

That's one lesson from and for the ESA. There's another, I believe.

If you should ever ask the folks in East Hokkaido, "So, are you going to keep feeding these birds every winter for eternity?" don't be too surprised if a number of them answer, "Well, sure, why not?"

The comedic author Bill Bryson once contrasted the difference between hiking along a nature trail in Europe versus the United States in one of his books. I can't recall which one, though. In the United States, he remarked on the wild abandon of the woods, how a hiker can very quickly find oneself in the middle of nowhere, with no services or signs of civilization to speak of, completely alone and in the elements. In Europe, by contrast, he said it was more of a mix: you're engulfed in nature and the wild at one moment, only to find yourself stumbling next to a rural cottage or encountering an abandoned road or railroad, perhaps with signs helpfully pointing you to services just a few hundred meters away. In the United States, there's this attempt to force a separation between humans and nature by an almost invisible and largely impenetrable wall. In Europe, however, nature and civilization seem to be more mixed or comingled, as Bryson describes.

The situation in Japan is more like the latter, and the Japanese are used to this. Many of them view wildlife almost like pets that are left to run loose. They think nothing of leaving food for them (with the glaring exception of bears) and seem comfortable when wild animals become almost domesticated in the human presence. In Nara, Japan's ancient capital, wild deer are ubiquitous and beg for food. They'll even eat snacks right out of your hand. Far from trying to stop this, the local park managers actually encourage it instead—deer feed is sold in vending machines, and visitors are allowed to feed the wildlife openly.

It's a different perspective, one that probably has something to do with the fact that Japan is overcrowded. The population here is about 125 million. Though the country is roughly the size of California, it's home to more than triple California's population. Conflicts with wildlife are inevitable. Accepting that their very existence is the cause of the demise of so many animals, the Japanese now accept it as their responsibility to ensure that the remaining wildlife survive and endure for successive generations to care for, like the Nara deer. And the serow. And the red-crowned crane. Perhaps one day we'll discover a way to de-extinct the Hokkaido wolf so that future generations of people on Hokkaido can have a chance to do better than their ancestors.

In the United States, we are in a post-Leopold world. In terms of parks management, the experts have already reached an agreement that the old ways of doing things are no longer adequate or even appropriate moving forward. We can't keep America's nature frozen in time as historical vignettes, windows into a pre-European-settlement past. This has been tried, and it's largely failed—the purple valleys of Rocky Mountain National Park are testament to how reality has a way of catching up with us, as a historically evergreen ecosystem becomes a mixed coniferous-deciduous one instead thanks to the mountain pine beetle. The nation's ever-swelling population, urban sprawl, habitat loss, and climate change make capturing and preserving natural America like a Polaroid picture an impossible task, a fool's errand. If this holds true for the national parks and the national wildlife refuges, then perhaps it holds true for America's endangered species at large, as well. This enforced hard separation between humans and threatened or endangered wildlife may no longer be possible.

All of the Earth is now spoken for, after all. The same holds for America—every square inch of it is spoken for—owned, traded, or set aside and managed to some degree, and the species residing on these remaining patches can no longer escape this reality, either. It may be time to figure out better ways of managing it all to the benefit of nature and humanity alike. A newer, more open-minded and experimental form of endangered species protections and recovery may be in order, and greater use of sustained, long-term artificial feeding in periods of food shortages for some critically endangered species, like the Attwater's prairie chicken, could play a part in a new management regime philosophy. Or not. After all, I'm not an expert on these matters, I'm just offering my two cents.

I return to the poem at the beginning of this book, and the hard truth it forces all of us to face:

> All are absorbed and reduced into individual molecules, tiny motes that show no hint of their vast history. In the bottomless sediments, only a vague memory remains.

One day, maybe sooner than I care to admit, the sea will reclaim me. And it will reclaim you. It will reclaim the whooping crane and the red-crowned crane, too. Thylacine may be revived from extinction soon. I earnestly hope so. But it will be driven to extinction again someday, unfortunately, despite our best efforts. Perhaps species managers will succeed in reviving the Attwater's prairie chicken. Perhaps the Mauritius parakeet will once again rule over that island. Maybe the white stork of Europe swells in population numbers so great that it spreads its wings to North America. But sometime, hopefully in the far distant future, all these species will go the way of the dinosaur. Humans will join them there eventually.

Evolution is amazing, truly "the greatest show on Earth," as famed biologist Richard Dawkins rightly observed. As slow as it is, under normal circumstances (meaning in the absence of a mass extinction event like the one humanity is causing now), evolution runs faster than extinction. We know this because the fossil record proves it. Still, extinctions are inevitable, as I've already acknowledged and grudgingly come to accept.

Something like 98 to 99 percent of all types of life that have ever existed on Earth are now extinct. For the vast majority of these cases, Homo sapiens had absolutely nothing to do with it. That's because our kind wasn't around when these extinctions happened, so we're blameless. Other now-extinct species may be to blame, but not us. But this has changed.

Today, there's good evidence that we are amid the sixth great mass extinction event, the Anthropocene extinction. This time, our species is the cause, without doubt. And though no other species on this planet will give a whit when our kind goes kaput for good (and humans will become extinct someday, I promise you), at least we have the capacity to think and act differently and even the tools and technologies at our disposal to prevent the extinctions of a host of other remarkable life-forms, all valuable in their own right. Perhaps, someday, we'll even have the technology to reverse extinctions, bringing back the Tasmanian tiger, the dodo, or maybe even the woolly mammoth.

Our kind is truly unique. We're the only species that cares whether other species live or die or go extinct entirely. We mourn the end of the dodo even to this day. Still, we have a lot more in common with the other members of the animal kingdom than you may think or are comfortable with accepting.

We know that animals have individual personalities and not just our domesticated pets—the phenomenon of individual animal personalities has been recognized and confirmed for wild animal species, as well. We know wild animals make friends. They feel all the same emotions you and I do. They cheat on their mates, and sometimes they get divorced. They love their offspring. And they mourn their dead. They have language. They use tools. They have culture even. Even plants may have far more interesting lives than we currently imagine. I've read studies that show how trees in a forest help each other survive droughts, sharing moisture and nutrients in times of need.

There are deeper similarities between humans and wildlife that I know you are almost certainly unaware of. I mention a few of them in my first book, *Anthill Economics: Animal Ecosystems and the Human Economy*.

I'm willing to bet that you live in a city, perhaps even a large one, or at least in some large population concentration. Do you want to know why we humans deliberately have chosen to crowd ourselves into our various metroplexes? Because that's what animals do. It's natural to do so, and nature has compelled us to do the same.

It's called "clustering" or sometimes "clumping" depending on which study you read, and it is far and away the most common form of animal population distribution seen on Earth. We didn't choose to urbanize; rather, we were compelled to do so by the same natural forces that drive the animal kingdom. And the effect is more than just "strength in numbers," as many of my students guess. It's more fundamental than that, which is why even some of the lowliest of species on this planet feel compelled to pile themselves into mass agglomerations of their own kind as their populations rise and rise, just as we did.

Here's more proof: falling birth rates.

As you and pretty much everyone else on the planet know by now, birth rates are falling nearly everywhere globally. In all heavily urbanized economies, reproduction rates are below what experts call the "replacement rate" necessary to sustain natural population growth. As a consequence, human population growth is slowing precipitously in a wide number of countries, while populations are actually declining in quite a few of them, as well, including Japan (though world population has reached eight billion and is destined to increase). This has been happening for some time, and now we see this phenomenon has caught up to the United States. The press calls this a disaster, a crisis. That smart space/electric car billionaire/Twitter owner says it's humanity's greatest threat. He's wrong. They're all wrong, in fact. The falling human fertility rate is perfectly natural, another sign that humanity is far more intimately intertwined with the natural order of things than we care to realize or admit.

Human birth rates are falling nearly everywhere because they *must* fall due to rising human population density. This same phenomenon has been witnessed time and time again for nearly every species of wildlife. The effect has even been confirmed in lab experiments. On the topic of wildlife population dynamics, naturalists know that, all things being equal, population growth tends to be higher when population density is low, and population growth tends to be lower when population density is high. That's what's happening to us—via the natural phenomenon of clustering, human population density is rising, compelling our species to lower our rate of reproduction due to the stress this crowding imposes upon us, the same thing that happens to numerous other species when their rates of population growth get out of control to the point at which their population densities rise and rise, leading to greater competition for space and resources.

I know of at least four peer-reviewed studies that have confirmed this interesting phenomenon: the higher a nation's population density, the lower its birth rate. Future studies will demonstrate the exact same thing. Population density is the strongest factor correlated with low and falling birth rates yet discovered, and that correlation grows stronger over time, according to the data. This implies causality. The latest study to confirm this was published in 2021.[18] It corresponds precisely with what's been witnessed and described in nature time and time again: as animal populations grow more densely populated, fertility rates decline.

How precisely the dynamics of this work, I won't go into here. I've already covered that in chapter 1 of my other book, *Anthill Economics*. But again, this phenomenon has been witnessed and described time and time again in the animal kingdom, causing me to conclude that birth rates are falling nearly everywhere because they *must* fall, because the laws of nature demand it for us, just as much as they do for our wild cousins in the animal kingdom with whom we share DNA. See? We have a lot more in common with animals than most of us care to admit.

Maybe somewhere deep down in this truth lies the fundamental reason why we're the only species on Earth that mourns the extinction of other species. Because, deep down, perhaps we instinctively know that we're related, even if we denied this fact for millennia. We understand that we're a part of nature; we feel it in our bones, even as we've done everything possible to separate ourselves from nature. We know that our rapacious demand for Earth's land and resources is killing off our neighbors en masse, causing a great extinction event to unfold right before our eyes. And we know that we can stop it, or at least that we are capable of putting a halt to it and perhaps even reversing it. So we try to do just that, and not just for two iconic crane species living on opposite ends of the world's largest ocean.

Why did we draft the Endangered Species Act, SARA, ACES, and all the other laws out there aimed at preventing what our ancestors would not? Because we care.

Is that enough?

~

Closing

On September 9, 2020, the World Wildlife Fund issued an alarming report. Following a global scientific study of the state of other living creatures on the planet and the extent of biodiversity loss experienced, WWF and its partners announced that scientists are confident that species populations globally have plummeted by 68 percent on average since 1970. Almost 70 percent fewer animals and plants alive on Earth since the Endangered Species Act became law. Not only are species being driven to extinction at alarming rates, but the numbers of animals and plants not yet rendered extinct have fallen precipitously, many of them to dangerously low levels. Populations of aquatic freshwater species are believed to have fallen by about 84 percent. Perhaps the emphasis on conservation spending on fish in the U.S. is justified. Biodiversity loss is the worst by far in Latin America and the Caribbean, where WWF and its partners say populations of living species have fallen by 94 percent on average. Now I know why I never encountered any wild animals, even birdsong, during my explorations of Colombia and Haiti. The report points to the usual culprits. "The most important direct driver for loss of biodiversity is land-use change, particularly the conversion of pristine native habitats, like forests, grasslands, and mangroves, into agricultural systems."[1] With the global human population at eight billion and rising, I fear this situation is about to get a lot worse.

Some of you reading this now already know about that WWF report and its alarming findings, even if a few years have passed since it was issued.

Most of you, however, never heard of it or had any inkling that biodiversity loss was this bad. This is probably because of the time when the report and WWF's findings were issued—toward the tail end of 2020 amid the COVID-19 pandemic and just two months before the U.S. presidential election. I didn't even know about this report until months after the news was issued. Back then, it was my job to pay attention to such developments, even to report on them, or so I thought. The press corps, or at least a majority of it, had other ideas. For them, there was only one story worth following and one single individual to whom to direct the bulk of our attention and coverage. That wildlife and wild plant populations have fallen globally by nearly 70 percent in just four decades didn't particularly warrant any serious consideration or attention, it was determined. Of course, some stories on the WWF report were almost certainly published or broadcasted. But far too few in my estimation. That's why we hardly noticed.

I began this book with some mild if pointed criticism of the media. We might as well end on this theme, too.

My critiques of the press corps are perhaps too harsh, but it's worthwhile to reiterate them since I experienced many of their failings firsthand. To be certain, there are scores of fabulous journalists out there who do tremendous jobs. I've had the pleasure of working alongside many of them (and still do so today). But the industry continues to be plagued with a terrible attention-span deficit that badly warps how we're currently seeing the world. I've never seen my own country more divided, and the media shares a huge portion of the blame for this state of affairs. The public is more than partly at fault here, too, but it's news professionals' jobs to rise above such obstacles to inform and educate the public to the best of their ability, not to entertain them or appeal to their basest desires. That's what we're setting out to do at PublicParks.org: a relentless focus on the facts, science, and policy behind the management of the world's public lands and waters. Sober and straightforward, but always accurate, informative, and educational, because that's what's the audience needs. They don't need clickbait and they don't require pandering to. Thus, my motto as far as news reporting is concerned: give the audience what it needs, not what it wants.

Reporters must resist the urge to descend to the level of clickbait and rage-porn, even if that's the direction commercial pressures are pushing them. This urge is ultimately self-defeating; it cannot be sustained. Worse, these trends are encouraging the news-consuming public of the United States and much of the world to split itself into warring camps—opposing tribes consuming only the infotainment they want while ignoring other events and other perspectives entirely. It's like we're seeing two different worlds, two different realties. This trend in modern news coverage also is causing the world to essentially overlook many of the most pressing crises of our age. Global warming receives plenty of media coverage, for sure, as it should. But the ongoing extinction crisis barely registers in the mass media these days. That could explain why the Recovering America's Wildlife Act still languishes in Congress as I type this. If reporters pay scant attention to it, then why should lawmakers bother? They can get away with ignoring it.

I searched "biodiversity" in Google News just now to see which media outlets are covering the crisis today. Here they are: the Nature Conservancy, Natural Resources Defense Council, Earth Justice, University of Connecticut, United Nations Conference on Trade and Development, Gov.UK, and Purdue University. Those, of course, aren't news outlets at all but mostly environmental nonprofits and universities. The *Financial Times* and *Eco-Business* have some items on how biodiversity is becoming a popular theme in environmental, social, and governance investing, or ESG, but that's pretty much it. *Nature* and Michigan State University had a couple items pop up when I did my search.

Will news outlets feel inspired to delve more deeply into this crisis on the occasion of the 50th anniversary of the U.S. Endangered Species Act? I don't recall that happening during the 40th anniversary, or the 30th or the 25th. Hopefully things will turn out differently this time around. But I won't hold my breath.

When a story goes untold, that gives policymakers license to ignore it. When a species goes extinct without making headlines, that gives readers and television viewers license to not care or to feel entirely indifferent about this loss. When plants and animals become shapes that we will never see again and we're indifferent to this, that gives license to the forces destroying biodiversity to continue going about their business, to persist in the destruction of 15 billion years' worth of slow creation without any repercussions whatsoever. Of course, when our kind goes extinct, no other creature on Earth will announce this to the world or shed a tear about this fact. We have the capacity to act and feel differently, however, even though we humans are also members of the animal kingdom.

A friend once asked on Facebook, and I'm paraphrasing here, "When humans are gone for good, will we have ever mattered at all?" It's a deeply philosophical question. Or maybe he was just drunk. Either way's good. I answered him this way, repeating something I've heard somewhere else that stuck with me: we are a means by which the universe can comprehend itself. That's something, I think.

Maybe we don't need a global solution to this global problem. Maybe local solutions will suffice. The first step to solving a crisis is knowing that it exists and how bad it is. That's the media's job. Helping our cousins in the animal kingdom to survive and thrive at least as long as our kind exists? That's on all of us.

~

Notes

Chapter 1

1. Musa Al-Ghabri, "The *New York Times*' Obsession with Trump, Quantified," *Columbia Journalism Review*, November 13, 2019.

2. International Union for Conservation of Nature, "Species Extinction—the Facts," IUCN Species Survival Commission, Geneva, May 2007.

Chapter 2

1. International Union for Conservation of Nature, "Species Extinction—the Facts," IUCN Species Survival Commission, Geneva, May 2007.

2. International Union for Conservation of Nature, "The Marseille Manifesto," IUCN World Conservation Congress, September 10, 2021.

3. IUCN, "The Marseille Manifesto."

Chapter 3

1. Nathanial Gronewold, "Comparative Conservation Strategy Efficacy for *Grus japonensis* and *Grus americana*: A Post-Policy Implementation Assessment," *Journal of International Wildlife Law & Policy* 24, no. 3–4 (2021): 224–50, doi: 10.1080/13880292.2021.2006422.

2. Lutz Tischendorf, "Population Viability and Critical Habitat in the Wood Buffalo National Park Area NT/AB Canada," National Wildlife Research Center, Canadian Wildlife Service, January 2003.

3. Carmen Davis, "A Review of the Success of Major Crane Conservation Techniques," *Bird Conservation International* 8, no. 1 (1998): 19–29, doi: 10.1017/S0959270900003609.

4. Matthew J. Butler, Kristine L. Metzger, and Grant Harris, "Whooping Crane Demographic Responses to Winter Drought Focus Conservation Strategies," *Biological Conservation* 179 (September 2014): 72–85.

5. R. Douglas Slack, William E. Grant, Stephen E. Davis III, Todd M. Swannack, Jeffrey Wozniak, Danielle M. Greer, and Amy G. Snelgrove, "San Antonio Guadalupe Estuarine System: Linking Fresh Water Inflows and Marsh Community Dynamics in San Antonio Bay to Whooping Cranes," Texas A&M AgriLife, April 2009.

6. Slack, Grant, Davis, Swannack, Wozniak, Greer, and Snelgrove, "San Antonio Guadalupe Estuarine System."

7. United States Fish & Wildlife Service, Aransas National Wildlife Refuge, Austwell, Texas, and Corpus Christi Ecological Service Field Office, Texas, "Whooping Crane (*Grus americana*) 5-Year Review: Summary and Evaluation," U.S. Fish & Wildlife Service, February 2012.

8. Guadalupe-Blanco River Authority and the Aransas Project, "GBRA-TAP Shared Vision for the Guadelupe River System and San Antonio Bay: Update on Implementation of the Agreement," January 2018.

9. Texas Parks & Wildlife Department, "Powderhorn Ranch Becomes Texas' Newest Wildlife Management Area," news release, October 25, 2018.

10. Jeffrey G. Paine, Edward W. Collins, Tiffany L. Caudle, and Lucie Costard, "Powderhorn Ranch Geoenvironmental Atlas," Bureau of Economic Geology, University of Texas at Austin, 2018.

11. "U.S. Army Corps of Engineers Coastal Texas Protection and Restoration Feasibility Study: Draft Integrated Feasibility Report and Environmental Impact Statement," U.S. Army Corps of Engineers, Galveston District, Southwest Division 2018.

12. D. Stratton, "Population Growth: The Return of the Whooping Crane," *Case Studies in Ecology and Evolution* (2010).

13. Environment Canada, "Recovery Strategy for the Whooping Crane (*Grus americana*) in Canada," Species at Risk Recovery Strategy series, Ottawa, 2007.

Chapter 4

1. Nathanial Gronewold, "NYC Brownfields Reclamation Effort Gets Off to Slow Start," *New York Times*, November 22, 2010.

2. Nathanial Gronewold, "Watchdogs Wearily Eye Rollout of New NJ Waste Site Cleanup Scheme," *New York Times*, May 26, 2009.

3. Pam LeBlanc, "Set to Become a State Park, Powderhorn Ranch Is an Unspoiled Coastal Paradise," *Texas Monthly*, February 7, 2022.

4. "Judge Orders Man to Pay $85K in Deaths of 2 Whooping Cranes," Associated Press, August 1, 2020.

5. Laura Schulte, "Shootings of 4 Rare Whooping Cranes in Oklahoma Raise Concerns about Proposed Wisconsin Sandhill Crane Hunt," *Milwaukee Journal Sentinel*, January 18, 2022.

6. Elisabeth Condon, William B. Brooks, Julie Langenberg, and Davin Lopez, "Whooping Crane Shootings since 1967," in *Whooping Cranes: Biology and Conservation*, ed. Philip J. Nyhus, John B. French, Sarah J. Converse, and Jane E. Austin (London: Elsevier, 2019), chap. 23.

7. William Hornaday, *Our Vanishing Wildlife: Its Extermination and Preservation* (New York: Charles Scribner's Sons, 1913).

8. Mose Buchele, "Biden Administration Considers Removing Whooping Cranes from the Endangered Species List," Texas Public Radio, December 13, 2021.

Chapter 5

1. Australian Science Media Centre, "Expert Reaction: The Iconic Koala Is Now Listed as Endangered," press release, February 11, 2022.

2. Australian Science Media Centre, "Expert Reaction."

3. Mingli Lin, Samuel T. Turvey, Chouting Han, Xiaoyu Huang, Antonios D. Mazaris, Mingming Liu, Heidi Ma, Zixin Yang, Xiaoming Tang, and Songhai Li, "Functional Extinction of Dugongs in China," *Royal Society Open Science* 9, no. 8 (August 24, 2022).

4. International Union for Conservation of Nature, press release, December 6, 2021.

5. William Hornaday, *Our Vanishing Wildlife: Its Extermination and Preservation* (New York: Charles Schribner's Sons, 1913).

6. Hornaday, *Our Vanishing Wildlife.*

7. Hornaday, *Our Vanishing Wildlife.*

8. Ministry of Land, Infrastructure and Transport, Hokkaido Regional Development Bureau, "Nature Restoration at the Kushiro Wetland," restoration planning overview, 2007.

9. Masashi Hanioka, Yuichi Yamaura, Masayuki Senzaki, Satoshi Yamanaka, Kazuhiro Kawamura, and Futoshi Nakamura, "Assessing the Landscape-Dependent Restoration Potential of Abandoned Farmland Using a Hierarchical Model of Bird Communities," *Agriculture, Ecosystems and Environment* 265 (June 2018).

10. Futoshi Nakamura, Tadashi Sudo, Satoshi Kameyama, and Mieko Jitsu, "Influences of Channelization on Discharge of Suspended Sediment and Wetland Vegetation in Kushiro Marsh, Northern Japan," *Geomorphology* 18, no. 3–4 (1997).

11. Nakamura, Sudo, Kameyama, and Jitsu, "Influences of Channelization."

12. Nakamura, Sudo, Kameyama, and Jitsu, "Influences of Channelization."

13. Shigeru Mizugaki, Futoshi Nakamura, and Tohru Araya, "Using Dendrogeomorphology and 137Cs and 210Pb Radiochronology to Estimate Recent Changes

in Sedimentation Rates in Kushiro Mire, Northern Japan, Resulting from Land Use Change and River Channelization," *CATENA* 68, no. 1 (2006).

14. Mizugaki, Nakamura, and Araya, "Using Dendrogeomorphology and 137Cs and 210Pb Radiochronology."

15. Futoshi Nakamura, Satoshi Kameyama, and Shigeru Mizugaki, "Rapid Shrinkage of Kushiro Mire, the Largest Mire in Japan, Due to Increased Sedimentation Associated with Land-Use Development in the Catchment," *CATENA* 55, no. 2 (2004).

16. Fujita Hiroko, "Outline of Mires in Hokkaido, Japan, and Their Ecosystem Conservation and Restoration," *Global Environmental Research* 11 (2007).

17. Yasunori Nakagawa, Takatoshi Nakamura, Hiroyuki Yamada, and Futoshi Nakamura, "Changes in Nitrogen and Base Cation Concentrations in Soil Water due to the Tree Cutting in a Wetland Alder Forest in the Kushiro Wetland, Northern Japan," *Limnology* 13 (2012).

18. Satoshi Yamanaka et al., "Influence of Farmland Abandonment on the Species Composition of Wetland Ground Beetles in Kushiro, Japan," *Agriculture, Ecosystems and Environment* 31 (2017).

19. Su Liying and Zou Hongfei, "Status, Threats and Conservation Needs for the Continental Population of the Red-Crowned Crane," *Chinese Birds* (September 2012).

20. Ho Gul Kim, Eun-jae Lee, Chan Park, Ki Sup Lee, Dong Kun Lee, Woo-shin Lee, and Jong-U Kim, "Modeling the Habitat of the Red-Crowned Crane (*Grus japonensis*) Wintering in Cheorwon-Gun to Support Decision Making," *Sustainability* 8, no. 6 (2016).

21. Liying and Hongfei, "Status, Threats and Conservation Needs."

22. Y. Masatomi, S. Higashi, and H. Masatomi, "A Simple Population Viability Analysis of Tancho (*Grus japonensis*) in Southeastern Hokkaido, Japan," *Population Ecology* 49 (2007).

23. Ministry of the Environment, "Red-Crowned Crane Habitat Expansion Strategic Plan (*tancho seisokuchi bunsan koudou keikaku*)," Hokkaido Regional Environment Office, Kushiro Branch, April 2013.

24. Ministry of the Environment, "Red-Crowned Crane Protection and Propagation, Plan Implementation Results Fiscal Year 2017 (*heisei 29 nendo tancho hogo zoushoku jigyou jisshi kekka*)," Hokkaido Regional Environment Office, Kushiro Branch, August 2018.

25. Keigo Nakamura, Klement Tockner, and Kunihiko Amano, "River and Wetland Restoration: Lessons from Japan," *BioScience* 56, no. 5 (May 2006).

26. Koichi Kuriyama, "Measuring the Ecological Value of the Forests around the Kushiro Marsh: An Empirical Study of Choice Experiments," *Journal of Forest Research* 5, no. 7–11 (2000).

27. Y. S. Ahn and F. Nakamura, "Landscape Restoration: A Case Practice of Kushiro Mire, Hokkaido," *Landscape Ecological Applications* (2008).

28. Keigo Nakamura, Klement Tockner, and Kunihiko Amano, "River and Wetland Restoration: Lessons from Japan," *BioScience* 56, no. 5 (May 2006); Nakajima,

Keiji, "National Report on the Implementation of the Ramsar Convention on Wetlands," Report to the 12th Meeting of the Conference on Contracting Parties, 2015.

29. Robert Tarrant and James Trappe, "The Role of *Alnus* in Improving the Forest Environment," *Plant and Soil* 35 (1971).

30. Carey Krajewski, "Phylogenetic Taxonomy of Cranes and the Evolutionary Origin of the Whooping Crane," in *Whooping Cranes: Biology and Conservation*, ed. Philip J. Nyhus, John B. French, Sarah J. Converse, and Jane E. Austin (London: Elsevier, 2019), chap. 2.

31. Aaron T. Pearse, Matt Rabbe, Lara M. Juliusson, Mark T. Bidwell, Lea Craig-Moore, David A. Brandt, and Wade Harrell, "Delineating and Identifying Long-Term Changes in the Whooping Crane (*Grus americana*) Migration Corridor," *PLOS ONE* 13, no. 2 (February 15, 2018).

32. Miranda R. Bertram, Gabriel L. Hamer, Karen F. Snowden, Barry K. Hartup, and Sarah A. Hamer, "Coccidian Parasites and Conservation Implications for the Endangered Whooping Crane (*Grus americana*)," *PLOS ONE* 10, no. 6 (June 10, 2015).

Chapter 6

1. Joel Methorst, Katrin Rehdanz, Thomas Mueller, Bernd Hansjürgens, Aletta Bonn, and Katrin Böhning-Gaese, "The Importance of Species Diversity for Human Wellbeing in Europe," *Ecological Economist* 181 (November 28, 2020).

2. Robin Waples, Marta Nammack, Jean Cochrane, and Jeffrey Hutchings, "A Tale of Two Acts: Endangered Species Listing Practices in Canada and the United States," *BioScience* 63, no. 9 (September 2013).

3. Center for Biological Diversity, "Lawsuit Launched to Protect Lesser Prairie Chicken from Extinction," press release, August 11, 2022.

4. Erin Ward, "The Gray Wolf under the Endangered Species Act (ESA): A Case Study in Listing and Delisting Challenges," Congressional Research Service (CRS) Report R46148, November 25, 2020.

Chapter 7

1. Katherine Gibbs and David Currie, "Protecting Endangered Species: Do the Main Legislative Tools Work?" *PLOS ONE* (May 2, 2012).

2. Gibbs and Currie, "Protecting Endangered Species."

3. Martin F. J. Taylor, Kieran F. Suckling, and Jeffrey J. Rachlinski, "The Effectiveness of the Endangered Species Act: A Quantitative Analysis," *BioScience* 55, no. 4 (April 2005).

4. Taylor, Suckling, and Rachlinski, "The Effectiveness of the Endangered Species Act."

5. Daniel M. Evans et al., "Species Recovery in the United States: Increasing the Effectiveness of the Endangered Species Act," *Issues in Ecology* 20 (Winter 2016).

6. Michael L. Casazza, Cory T. Overton, Thuy-Vy D. Bui, Joshua M. Hull, Joy D. Albertson, Valary K. Bloom, Steven Bobzien, Jennifer McBroom, Marilyn Latta, Peggy Olofson, Tobias M. Rohmer, Steven E. Schwarzbach, Donald R. Strong, Erik Grijalva, Julian K. Wood, Shannon Skalos, and John Y. Takekawa, "Endangered Species Management and Ecosystem Restoration: Finding the Common Ground," *Ecology and Society* 21, no. 1 (March 2016).

7. Daniel M. Evans, Judy P. Che-Castaldo, Deborah Crouse, Frank W. Davis, Rebecca Epanchin-Niell, Curtis H. Flather, R. Kipp Frohlich, Dale D. Goble, Ya-Wei Li, Timothy D. Male, Lawrence L. Master, Matthew P. Moskwik, Maile C. Neel, Barry R. Noon, Camille Parmesan, Mark W. Schwartz, J. Michael Scott, and Byron K. Williams, "Species Recovery in the United States: Increasing the Effectiveness of the Endangered Species Act," *Issues in Ecology* 20 (Winter 2016).

8. Euan Ritchie, Deakin University, Australian Science Media Centre, press release, December 20, 2022.

9. International Union for Conservation of Nature, "European Bison Recovering, 31 Species Declared Extinct—IUCN Red List," press release, December 20, 2020.

10. Ismael Soto, Ross N. Cuthbert, Antonin Kouba, Cesar Capinha, Anna Turbelin, Emma J. Hudgins, Christophe Diagne, Franck Courchamp, and Phillip J. Haubrock, "Global Economic Costs of Herpetofauna Invasions," *Scientific Reports* 12 (July 28, 2022).

11. Kurt E. Ingeman, Lily Z. Zhao, Christopher Wolf, David R. Williams, Amelia L. Ritger, William J. Ripple, Kai L. Kopecky, Erin M. Dillon, Bartholomew P. DiFiore, Joseph S. Curtis, Samantha R. Csik, An Bui, and Adrian C. Stier, "Glimmers of Hope in Large Carnivores' Recoveries," *Scientific Reports* 12 (July 21, 2022).

Chapter 8

1. "Red-Crowned Crane Feeding Volume Reductions to Be Eased," *Hokkaido Shimbun*, September 10, 2020.

2. "Revision to Crane Feeding Reductions in Tsurui," *Kushiro Shimbun*, November 9, 2021.

3. Kurt Ingeman et al., "Glimmers of Hope in Large Carnivore Recoveries," *Scientific Reports* 12 (July 2022).

Chapter 9

1. R. Hilgartner, D. Stahl, and D. Zinner, "Impact of Supplementary Feeding on Reproductive Success of White Storks," *PLOS ONE* 9, no. 8 (August 13, 2014).

2. Hilgartner, Stahl, and Zinner, "Impact of Supplementary Feeding."

3. Hilgartner, Stahl, and Zinner, "Impact of Supplementary Feeding."

4. Simon Tollington, Andrew Greenwood, Carl G. Jones, Paquita Hoeck, Aurélie Chowrimootoo, Donal Smith, Heather Richards, Vikash Tatayah, and Jim J. Groombridge, "Detailed Monitoring of a Small but Recovering Population Reveals Sublethal Effects of Disease and Unexpected Interactions with Supplemental Feeding," *Journal of Animal Ecology* 84 (2015): 969–77.

5. Tollington et al., "Detailed Monitoring."

6. Ainara Cortes-Avizanda, Guillermo Blanco, Travis L. DeVault, Anil Markandya, Munir Z. Virani, Joseph Brandt, and José A. Donázar, "Supplementary Feeding and Endangered Avian Scavengers: Benefits, Caveats, and Controversies," *Frontiers in Ecology and the Environment* 14, no. 4 (May 2016).

7. Miguel Ferrer, Ian Newton, Roberto Muriel, Gerardo Báguena, Javier Bustamante, Matilde Martini, and Virginia Morandini, "Using Manipulation of Density-Dependent Fecundity to Recover an Endangered Species: The Bearded Vulture *Gypaetus barbatus* as an Example," *Journal of Applied Ecology* 51 (2014).

8. Pavla Hejcmanová, Pavla Vymyslická, Magdalena Žáčková, and Michael Hejcman, "Does Supplemental Feeding Affect Behavior and Foraging of Critically Endangered Western Giant Eland in an ex situ Conservation Site?" *African Zoology* 48 (October 2013).

9. Daniel M. Evans, Judy P. Che-Castaldo, Deborah Crouse, Frank W. Davis, Rebecca Epanchin-Niell, Curtis H. Flather, R. Kipp Frohlich, Dale D. Goble, Ya-Wei Li, Timothy D. Male, Lawrence L. Master, Matthew P. Moskwik, Maile C. Neel, Barry R. Noon, Camille Parmesan, Mark W. Schwartz, J. Michael Scott, and Byron K. Williams, "Species Recovery in the United States: Increasing the Effectiveness of the Endangered Species Act," *Issues in Ecology* 20 (Winter 2016).

10. Evans et al., "Species Recovery in the United States."

11. Evans et al., "Species Recovery in the United States."

12. Kurt Ingeman, Lily Zhao, Christopher Wolf, David Williams, Amelia Ritger, William Ripple, Kai Kopecky, Erin Dillon, Bartholomew DiFiore, Joseph Curtis, Samantha Csik, An Bui, and Adrien Stier, "Glimmers of Hope in Large Carnivore Recoveries," *Scientific Reports* 12 (July 2022).

Chapter 10

1. Neil deGrasse Tyson and Donald Goldsmith, *Origins: Fourteen Billion Years of Cosmic Evolution* (New York: W. W. Norton, 2004).

2. Brian Greene, *Fabric of the Cosmos: Space, Time, and the Texture of Reality* (New York: 2005 Vintage, 2005).

3. Tyson and Goldsmith, *Origins*, 26.

4. Bill Bryson, *A Short History of Nearly Everything* (New York: Broadway Books, 2004).

5. Jan Borgelt, Martin Dorber, Marthe Alnes Høiberg, and Francesca Verones, "More Than Half of Data Deficient Species Predicted to Be Threatened by Extinction," *Communications Biology* 5, no. 679 (2022): 5.

6. Borgelt, Dorber, Høiberg, and Verones, "More Than Half."

7. Hiroki Teraoka, Erika Okamoto, Moe Kudo, Shouta M. M. Nakayama, Yoshinori Ikenaka, Mayumi Ishizuka, Tetsuya Endo, Takio Kitazawa, and Takeo Hiraga, "Accumulation Properties of Inorganic Mercury and Organic Mercury in the Red-Crowned Crane *Grus Japonensis* in East Hokkaido, Japan," *Ecotoxicology and Environmental Safety* 122 (2015).

8. Kensaku Kakimoto, Kazuhiko Akutsu, Haruna Nagayoshi, Yoshimasa Konishi, Keiji Kajimura, Naomi Tsukue, Tomoo Yoshino, Fumio Matsumoto, Takeshi Nakano, Ning Tang, Kazuichi Hayakawa, and Akira Toriba, "Persistent Organic Pollutants in Red-Crowned Cranes (*Grus japonensis*) from Hokkaido, Japan," *Ecotoxicology and Environmental Safety* 147 (2018).

9. Craig Hebert, "The River Runs through It: The Athabasca River Delivers Mercury to Aquatic Birds Breeding Far Downstream," *PLOS ONE* (April 9, 2019).

10. Hebert, "The River Runs through It."

11. "Red-Crowned Crane, a National Treasure, Dies after Being Shot with an Air Rifle in a Field—Ikeda," *Tokachi Mainichi News*, June 6, 2021.

12. Elisabeth Condon, William B. Brooks, Julie Langenberg, and Davin Lopez, "Whooping Crane Shootings since 1967," in *Whooping Cranes: Biology and Conservation*, ed. Philip J. Nyhus, John B. French, Sarah J. Converse, and Jane E. Austin (London: Elsevier, 2019), chap. 23.

13. Jeb Barzen and Ken Ballinger, "Sandhill and Whooping Cranes," Wildlife Damage Management Technical Series, U.S. Department of Agriculture Animal and Plant Health Inspection Service, January 2017.

14. Eric K. Waller, Theresa M. Crimmins, Jessica J. Walker, Erin E. Posthumus, and Jake F. Weltzin, "Differential Changes in the Onset of Spring across U.S. National Wildlife Refuges and North American Migratory Bird Flyways," *PLOS ONE* 13, no. 9 (September 12, 2018).

15. Waller, Crimmins, Walker, Posthumus, and Weltzin, "Differential Changes."

16. Joseph Chamie, "After Years of U.S. Population Growth, It's Time for a Pause," *The Hill*, March 7, 2022.

17. Kylee A. Pawluk, Caroline H. Fox, Christina N. Service, Eva H. Stredulinsky, and Heather M. Bryan, "Raising the Bar: Recovery Ambition for Species at Risk in Canada and the U.S.," *PLOS ONE* 14, no. 11 (November 19, 2019).

18. Amanda Rotella, Michael E. W. Varnum, Oliver Sng, and Igor Grossman, "Increasing Population Densities Predict Decreasing Fertility Rates over Time: A 174-Nation Investigation," *American Psychologist* 76, no. 6 (2021), doi: https://doi.org/10.1037/amp0000862.

Closing

1. World Wildlife Fund, "68% Average Decline in Species Population Sizes since 1970, Says New WWF Report," press release, September 9, 2020.

~

Index